COMPUTERS

Recent Titles in
Greenwood Technographies

Sound Recording: The Life Story of a Technology
David L. Morton Jr.

Rockets and Missiles: The Life Story of a Technology
A. Bowdoin Van Riper

Firearms: The Life Story of a Technology
Roger Pauly

Cars and Culture: The Life Story of a Technology
Rudi Volti

Electronics: The Life Story of a Technology
David L. Morton Jr. and Joseph Gabriel

The Railroad: The Life Story of a Technology
H. Roger Grant

COMPUTERS

THE LIFE STORY OF A TECHNOLOGY

Eric G. Swedin and David L. Ferro

GREENWOOD TECHNOGRAPHIES

GREENWOOD PRESS
Westport, Connecticut • London

Library of Congress Cataloging-in-Publication Data

Swedin, Eric Gottfrid.
Computers : the life story of a technology / Eric G. Swedin and David L. Ferro.
p. cm.—(Greenwood technographies, ISSN 1549–7321)
Includes bibliographical references and index.
ISBN 0–313–33149–9 (alk. paper)
1. Computers—History. I. Ferro, David L. II. Title. III. Series.
QA76.17.S94 2005
004—dc22 2004028174

British Library Cataloguing in Publication Data is available.

Library of Congress Catalog Card Number: 2004028174
ISBN: 0–313–33149–9
ISSN: 1549–7321

First published in 2005

Greenwood Press, 88 Post Road West, Westport, CT 06881
An imprint of Greenwood Publishing Group, Inc.
www.greenwood.com

Printed in the United States of America

The paper used in this book complies with the Permanent Paper Standard issued by the National Information Standards Organization (Z39.48–1984).

10 9 8 7 6 5 4 3 2 1

Contents

Series Foreword

In today's world, technology plays an integral role in the daily life of people of all ages. It affects where we live, how we work, how we interact with each other, and what we aspire to accomplish. To help students and the general public better understand how technology and society interact, Greenwood has developed *Greenwood Technographies*, a series of short, accessible books that trace the histories of these technologies while documenting *how* these technologies have become so vital to our lives.

Each volume of the *Greenwood Technographies* series tells the biography or "life story" of a particularly important technology. Each life story traces the technology from its "ancestors" (or antecedent technologies), through its early years (either its invention or development) and rise to prominence, to its final decline, obsolescence, or ubiquity. Just as a good biography combines an analysis of an individual's personal life with a description of the subject's impact on the broader world, each volume in the *Greenwood Technographies* series combines a discussion of technical developments with a description of the technology's effect on the broader fabric of society and culture—and vice versa. The technologies covered in the series run the gamut from those that have been around for centuries—firearms and the printed book, for example—to recent inventions that have rapidly taken over the modern world, such as electronics and the computer.

While the emphasis is on a factual discussion of the development of the technology, these books are also fun to read. The history of technology is full of fascinating tales that both entertain and illuminate. The authors—all experts in their fields—make the life story of technology come alive, while also providing readers with a profound understanding of the relationship of science, technology, and society.

Introduction

Computers are the great technological and scientific innovation of the last half of the twentieth century. The computer has changed how we work, how we organize and store information, how we communicate with each other, and even the way that we think about the universe and the human mind. Computers have alleviated the drudgery of calculating sums and clerical work, and become essential tools in all facets of technological industries and everyday life. Computers have become ubiquitous in many aspects of business, recreation, and everyday life, and the trend is that they are becoming more powerful, more commonplace, and easier to use. *Computers: The Life Story of a Technology* tells the story of this evolution.

The story of the computer did not begin in the twentieth century. Many ancient civilizations sought ways to automate mathematics. The clay tablets of Babylon, the Roman and Chinese abacuses, the mechanical adders of Pascal in the 1640s, and the steam-powered devices imagined by Charles Babbage in the nineteenth century all led toward the modern computer.

World War II provided the impetus for the development of the electronic digital computer. After the war, the Cold War security and defense needs of the United States drove the development of computing technology. These advances occurred principally in the United States: the "giant brains" of the SAGE early warning system included an interactive interface, the connecting of computers across the country in ARPANET led to the Internet, and the constant miniaturization of circuitry for use in space and missile technology depended on the creation of integrated circuits.

Hundreds of millions of computers around the world serve us in many ways, from helping us to write books to microwaving our food. Only a couple decades ago most people couldn't imagine the value of a computer in their homes, while today a majority of households in Western society have more than one. In the workplace, computers have changed the lay of the corporate landscape. Computers have made clerical work more efficient, and have raised the expectations of greater individual productivity. The role of the secretary has been reduced since individuals within the workplace are now expected to master word processing, spreadsheet, database, and numerous other programs on their personal workstations. The use of these programs has increased the expected turnaround times for any task that these programs facilitate. This book traces the effect of automation on the workplace: from the work of Herman Hollerith on the 1890 U.S. Census, through the efforts of IBM in the 1950s, to the advent of the microcomputer in the 1970s.

The computer created a nexus through which two major trends in human development—advances in communication and automatic calculation—came together. With the development of digital circuitry we see the digitization of the senses: motion, sound, the written word, and even tastes and smells, given the right technology. With the advent of the different networks that make up the Internet, we see the possibility of vast volumes of digitized content moving across the globe in milliseconds. With networked microcomputers, including smart phones, storing this information has became decentralized. The ease of manipulating digital content either purposefully or accidentally—especially text and pictures—lends itself to fraud. As human production is digitized, identification and authentication practices struggle to catch up.

The computer has become such a powerful device that it lends itself as a powerful metaphor as well. Much as the clock in the Middle Ages changed the way that people interacted with reality through time, and Newtonian physics and the invention of the steam engine in the eighteenth century stimulated scientists to think of the laws of nature in terms of machines, the success of the computer in the late twentieth century prompted scientists to think of the basic laws of the universe as being similar to the operation of a computer. The new physics of information has come to view matter and natural laws as bits of information. So too did the computer change our way of thinking about . . . thinking. Through their efforts to create artificial intelligence, scientists reimagined the mind in terms of computer resources and discovered new insights into the biological mechanisms of thought and memory.

Computers: The Life Story of a Technology provides an accessible historical overview of this ever-changing technology, giving students and lay readers an understanding of the scope of its history from ancient times to the present day. It illuminates the details of the technology while also linking those developments to the historical context of the times. This book is about the story of computers, but is also the story of the people and events that drove the many technological innovations that led to modern electronic computers. Both of the authors have spent over two decades in the computer field and have watched history unfold. We began our careers when punched card readers were still used and are still actively engaged in our exciting field.

Timeline

35,000 BCE	Notched bones used for counting.
20,000 BCE	Pictures on rocks appeared in Europe.
10,000 BCE	Clay objects for counting used in the Fertile Crescent.
6000 BCE	First seals that make impressions on clays.
4000 BCE	People of Sumer invent a base 60 numbering system (the base of the time) that is later used by Babylonians, Greeks, and Arabs.
3300 BCE	Sumerian written number system.
3000 BCE	Egyptian number system and hieroglyphics appear.
2700 BCE	An abacus invented in Sumer and Babylonia.
2000 BCE	Minoan civilization uses tablets of clay for accounting calculations.
1500 BCE	The first alphabetic language appears with the Semites in what today is Syria.
1000 BCE	Many numeric systems appear.
300 BCE	First known use of zero by Babylonian scholars.
200 BCE	Reference to both sand and token types of abacus in Greece. Chinese invent paper and the mambo stick abacus.

100 BCE	Romans are using the wax abacus and the abacus with rods.
0 CE	Romans are using the pocket abacus.
300	Mayans are using counting for astronomy and later develop a base 20 number system with zero.
500	A nine-digit system invented in India is derived from Brahmi notation, where the number's positions have significance. The Indians begin to move beyond column significance to location from decimal and fully develop an understanding of the positional use of zero.
800	The concept of the zero makes its way to China and Islam.
900	"Arabic" numbers and the zero are introduced to Western Europe through Spain.
1000–1100	Block printing is invented by the Chinese.
1200	Widespread use of Indian digits 0 through 9, which are called "algorisms" in Europe.
1202	The Italian Leonardo of Pisa (known as Fibonacci) publishes *Liber Abaci* (A Treatise on the Abacus) from where many algebraic techniques come, including the Fibonacci sequence.
1300	Chinese are using the Chinese abacus.
1500–1540	Printing reinvented in Europe by Johannes Gensfleisch (known as Gutenberg).
1600–1614	Scotsman John Napier invents a way to calculate logarithms using "Napier's Bones"—ivory rods that work as a calculator when appropriately arranged.
1622	First slide rule invented by William Oughtred.
1637	René Descartes invents analytical geometry.
1642	Pascal designed a mechanical calculator (addition only) called the Pascaline.
1680s	Gottfried Wilhelm Leibniz and Isaac Newton each independently invent calculus.
1694	Leibniz creates the Leibniz wheel, which is a four-function device (addition, subtraction, multiplication, and division), saving much time for creating logarithm tables.
1700s	The rise of increasingly accurate analog mechanical devices.

1800–1801	Joseph Jacquard creates and implements memory and programmability by using a punched card on a loom.
1811	Luddites, fearing loss of their livelihood, try to destroy many Jacquard looms.
1820	The Arithmometer, the first mass-produced calculator, introduced.
1823	Charles Babbage creates his Difference Engine, which can add, subtract, multiply, and divide to six significant figures, but he never attempts the twenty significant figures needed for real usefulness.
1830s	Babbage designs his Analytic Engine, which functions similar to a modern digital computer with a processor, storage for data, and cards for input and output. He fails to secure enough funding to build thc machine and dies before trying to build the machine.
1842	Ada Lovelace becomes the first programmer by describing Babbage's Analytic Engine.
1847–1854	George Boole creates boolean algebra and publishes *An Investigation of the Laws of Thought.*
1868	The QWERTY keyboard is invented by Christopher Sholes, specifically designed to slow down typists so that the mechanical keys on a typewriter will not hit each other.
1888	William S. Burroughs patents a mechanical adding machine.
1890	Herman Hollerith patents a set of three machines, including a mechanically programmable enumeration machine (talley or sort) using keypunched cards, called a Tabulating Machine. Hollerith machines are used for the 1890 U.S. Census.
1896	Hollerith establishes the Tabulating Machine Company, which later becomes International Business Machines (IBM).
1904	John Fleming invents the diode vacuum tube, which can convert alternating current (AC) to direct current (DC).
1910–1915	The physicist Manson Benedicks realizes that a germanium crystal can convert AC to DC, which later becomes a basic building block for microchips.
1919	The flip-flop switch is invented by W. H. Eccles and F. W. Jordan.
1920–1921	The Czech playwright Karel Cápek coins the word "robot" in his play *RUR* (*Rossum's Universal Robots*).

1924	Thomas John Watson becomes the chairman of the newly renamed International Business Machines (IBM).
1927	Television is demonstrated to the public with the broadcast of a speech in Washington, D.C., sent to New York City.
1929	Bell Labs demonstrates the first color television.
1935	IBM introduces the electric typewriter and the 601 punch card machine and trains a class of women as service technicians.
1937	A early version of the binary adder circuit is created by George Stibitz. The "Turing machine," a theoretical model of a modern computer, is proposed by the mathematician Alan Turing in England.
1938	Hewlett-Packard Company founded by Bill Hewlett and Dave Packard to create specialized calculating equipment. The German Konrad Zuse builds his Z1, an electromechanical computer.
1939–1942	John Vincent Atanasoff and Clifford E. Berry at Iowa State University create a linear equation computer (later called the ABC computer), based on vacuum tubes, with in-memory programs.
1943	The all-electronic computer Colossus is invented in England for use in codebreaking. The Navy-funded Mark I calculating machine is built at Harvard University under the direction of Howard Aiken at Harvard University. While working on the proposal for his machine, Aiken discovers that Harvard actually owned a piece of Babbage's Analytic Engine and is inspired by Babbage's earlier work. Grace Murray Hopper later becomes a programmer on the Mark I.
1945	The Army-funded Electronic Numerical Integrator and Computer (ENIAC) is built at the University of Pennsylvania Moore School of Electrical Engineering to calculate artillery ballistic tables. J. Presper Eckert and John Mauchly are the main engineers on the project. The ENIAC contains over 18,000 vacuum tubes, and is programmed by women using over 6,000 switches and plugs. It is a digital electronic computer with no way to store a program. Eckert and Mauchly begin work on the Electronic Discrete Variable Automatic Computer (EDVAC). Their work is adapted and expanded by the mathematician John von Neumann, who outlines the stored program concept in an influential paper and becomes known as the "father of computing."

Vannevar Bush publishes the seminal 1945 article, "As We May Think," in which he envisions the use of computers to organize information in a linked manner that we now recognize as an early vision of hypertext.

1947 The first transistor is invented at Bell Labs.

1948 Magnetic drum memory is developed for mass storage.
The Electronic Delay Storage Automatic Calculator (EDSAC) is built at the University of Manchester, a prototype that demonstrates von Neumann's ideas of stored variables and programs in electronic memory.

1950 Alan Turing posits the "Turing Test," a way of deciding if machines truly "think." It involves having a user interact with a machine and not detecting that it is a machine.

1951 J. Presper Eckert and John Mauchly build the first commercial computer, the Universal Automatic Computer (UNIVAC).

1952 The first UNIVAC is used to predict the victory of Dwight D. Eisenhower in the 1952 presidential election.

1953 IBM introduces the IBM 650 and IBM 701. The 650, while more limited than the UNIVAC, benefits from IBM's relationships with businesses and becomes the best-selling computer of the 1950s.
Magnetic core memory is perfected by Jay Forrester for use on the Air Force–funded SAGE computers. It solves the problem of the complicated and temporary memory-storing techniques used earlier. The first higher level programming language—Short Order Code—is invented by John Mauchly for use on the UNIVAC.

1954 A 600-line-per-minute printer called the Uniprinter is invented by Earl Masterson at Univac.

1956 Random access hard drives invented at IBM.
John McCarthy and Marvin Lee Minsky organize a summer seminar at Dartmouth College on artificial intelligence.

1957 John W. Backus develops the FORTRAN language and compiler.

1958 Digital Equipment Corporation is founded by Kenneth Olsen.
The modem, a device allowing digital signals to be transmitted through analog phone lines, is created at Bell Labs.
LISP (LISt Processing), the first nonprocedural language devoted to artificial intelligence research, is invented by John McCarthy on an IBM 704 at MIT.

1958–1959 Integrated circuits are independently invented by Jack S. Kilby of Texas Instruments and Robert Noyce at Fairchild Semiconductor in 1958 and 1959.

1959 The first copy machine is released by Xerox.
General Electric creates a machine that can read magnetic ink.

1960 Digital Equipment Corporation (DEC) premiers its PDP-1, the first minicomputer.
Grace Murray Hopper helps develop Common Business-Oriented Language (COBOL).

1961 Georg Devol patents the Unimate, an industrial robot. IBM introduces the 7030.

1962 Purdue University and Stanford University create the first computer science departments.
The first computer game—Spacewar—is invented by MIT grad student Steve Russell.

1963 The deployment of the Semi-automatic Ground Environment (SAGE) system is completed, building on the developments that began with Project Whirlwind in the late 1940s. It is a real-time system for U.S. national air defense.
The American Standard Code for Information Interchange (ASCII), a character code for character representation in computers, is developed.

1964 Thomas Kurtz and John Kemeny develop BASIC programming at Dartmouth College.
Control Data Corporation (CDC) brings the supercomputer to the marketplace, a CDC 6600 invented by Seymour Cray. Cray later founds Cray Research to build supercomputers.
The first computer-aided design (CAD) system is developed at IBM.
The computer mouse is invented at SRI by Douglas Engelbart.

1965 J. A. Robinson develops unification; important for logic programming.

1967 The first object-oriented programming language—Simula—is developed by Ole-Johan Dahl and Kristen Nygaard for use in creating airplane simulations.
Texas Instruments releases a handheld calculator that can add, subtract, multiply, and divide.

1968 The term "software engineering" is coined in a NATO science committee meeting. The YYMMDD date standard is set by the

	Federal Information Processing Committee—this will haunt programmers years later with the approach of the year 2000 and the "Y2K bug." Intel is founded by Gordon Moore (of the famous Moore's law), Andy Grove, and Robert Noyce.
1969	ARPANET, consisting of four nodes, comes online, eventually spawning the Internet. Neil Armstrong walks on the moon.
1970	Kenneth Thompson and Dennis Ritchie create the UNIX operating system at Bell Labs on a DEC PDP-7. The 8-inch floppy is introduced by IBM. Niklaus Wirth creates the Pascal programming language.
1971	First use of the "@" sign for an electronic message sent by Ray Tomlinson through the ARPANET. Intel creates the first microprocessor (a computer on a single chip) for use in a calculator.
1972	Hewlett-Packard introduces the replacement for the slide rule—the first handheld scientific calculator. Pong, the first video game to stand alongside pinball machines, is introduced by Atari, a company founded by Nolan Bushnell.
1973	The ENIAC patent is invalidated and the federal government recognizes John Atanasoff as the modern computer's inventor with his ABC computer design.
1974	A WYSIWYG (what you see is what you get) program called Bravo is introduced by Charles Simonyi at Xerox PARC. A program named Kaissa wins the first world computer chess tournament.
1975	Laser printing introduced by IBM. Ed Roberts offers the Altair 8800, an electronics kit to build a personal computer, for sale. Microsoft (first called Micro-Soft) is formed by Bill Gates and Paul Allen to sell their BASIC interpreter on the Altair 8800.
1976	Steve Wozniak and Steve Jobs create the Apple I microcomputer. It is an instant hit with hobbyists. The company OnTyme introduces the first commercial e-mail service. Proof of the four-color theorem is published, which marks the first time that a computer had been used to construct a formal mathematical proof. The public key Data Encryption Standard (DES) is released.

1977	Apple releases the Apple II, a computer that promises to work "right out of the box." The Apple II is a commercial hit. The PET microcomputer is released by Commodore.
1978	The Wordstar word processing program is released. DEC releases the VAX 11/780, a 32-bit minicomputer, with the VMS operating system. Epson releases a successful dot matrix printer.
1979	Dan Bricklin and Bob Frankston create the first electronic spreadsheet, VisiCalc, in Frankston's attic on an Apple II. It becomes the first "killer app," an application that drives hardware sales. The 16-bit 68000 microprocessor is released by Motorola. Cellular phones are first tested.
1980	IBM begins development of the IBM PC, choosing Microsoft's PC-DOS over Digital Research's CP/—86 as the operating system. The programming language Ada is released on the anniversary of Ada Lovelace's birthday, December 10. It is touted as the language in which all U.S. Defense Department programs will be written in the future, though it never quite achieves that. The 1980s' most successful database product for the PC debuts: dBase II, written by Wayne Ratliff.
1981	First space shuttle launched. The IBM Personal Computer (PC) is brought to market. The open architecture of the system leads to PC "clones" in the following few years. The first clone is from Columbia Data Products in 1982. Soon after, Compaq becomes the biggest competitor in the PC clone market. Steve Jobs visits the Xerox Palo Alto Research Center (PARC) and sees their inventions of the past decade: a graphical user interface using bitmapped graphics, menus, icons, and the mouse; networked computers; and graphical word processing and desktop publishing.
1982	*Time* magazine makes the personal computer its "Man of the Year." A number of cities in the United States now have commercial e-mail available. Adobe Systems, founded by John Warnock and Charles Geschke, creates the Postscript printing language. Autodesk creates AutoCAD. Intel releases the 16-bit 80286 microprocessor, eventually found in tens of millions of PCs.
1983	The first killer app for the IBM PC, Lotus 1-2-3, a spreadsheet, is brought to market. The Internet protocol TCP/IP becomes standard for the Internet. C++ is developed at Bell Labs by Bjarne Stroustrup.

	The Apple Lisa microcomputer is released and fails in the marketplace.
1984	In a commercial during the Superbowl, using Orwellian imagery, Apple introduces the Macintosh, a successor to the Lisa, the first affordable microcomputer with a graphical user interface (GUI) and mouse. The CD-ROM is introduced by Sony and Philips. Hollywood begins to revolutionize movie special effects with the use of computer graphics—notably in *The Last Starfighter.* William Gibson coins "cyberspace" in his novel *Neuromancer.*
1985	Windows 1.0, Microsoft's answer to the Macintosh, is released. The 80386 microprocessor is released by Intel. The field of PC desktop publishing is established with the release of Paul Brainard's PageMaker. Two machines, the Cray 2 and the Connection Machine (a parallel processing computer from Thinking Machines), achieve 1 billion operations per second. The National Science Foundation establishes four supercomputer centers connected to the Internet.
1986	The Compaq Deskpro 386 is the first PC to use the new 32-bit Intel 80386 microprocessor. This is a key turning point, when IBM begins to lose control of the PC architecture that they had created, since Compaq brought their PC to market first, not waiting to see what IBM would do.
1987	The Apple Macintosh II is released, and IBM creates a new generation of personal computers called the PS/2.
1988	The Morris worm, written by Robert Morris at Cornell University, brings down a quarter of the Internet. Motorola's 88000 microprocessor can process 17 million instructions per second.
1989	Tim Berners-Lee proposes the World Wide Web to his employer, Conseil Européen pour la Recherche Nucléaire (CERN), for the sharing of scientific information using hypermedia and the Internet. Intel releases the 80486 microprocessor, a microchip containing 1.2 million transistors.
1990	Microsoft Windows 3.0 launched.
1991	Cold War ends. Tim Berners-Lee releases the software for the World Wide Web with a web server and web client using the HTTP, URL, and HTML protocols.

1992	Windows 3.1 launched. IBM is no longer the largest seller of microcomputers (or PCs, a name they made synonymous with microcomputers).
1993	Intel releases the first Pentium microprocessor. The Newton, a personal digital assistant (PDA), is released by Apple, though it fails in the marketplace due to the perception that it suffers from poor handwriting recognition. NCSA Mosaic, the first graphical web browser, is released for free.
1994	Marc Andreesen and Jim Clark popularize web surfing considerably with the release of the Netscape web browser.
1995	Microsoft Windows 95 released. Sun Microsystems releases an object-oriented language named Java that promises to be platform independent. *Toy Story*, the first full-length computer-generated feature film, is released by Pixar Animation Studios, a company headed by Steve Jobs. Amazon.com and eBay.com open for business.
1996	The Palm Pilot, the first successful and affordable personal digital assistant (PDA), is released by 3Com.
1997	IBM's Deep Blue defeats the world chess-playing champion Garry Kasparov.
1998	Microsoft's Windows 98 is released.
1999	World population passes 6 billion humans. America Online (AOL) reaches millions of subscribers, helping to expand Internet use. Napster, a file-trading service on the Internet founded by Shawn Fanning, a student at Northeastern University, leads to a debate over music copyrights and prosecution by the music recording industry.
2000	The much-hyped Year 2000 (Y2K) bug does not lead to major power outages and crashing airliners. The problem was successfully contained by a rush of programming fixes in the few years preceding the century mark. The dot-com stock "bubble" lives up to its name as the virtual real estate grab by dot-com companies proves unsustainable. Microsoft is judged a monopoly by a federal judge, but little comes of it in years to come.
2001	Dell Computer becomes the largest seller of personal computers.
2002	Earth Simulator, a supercomputer from Hitachi, runs at 40 trillion operations per second.

	The .NET development environment is released by Microsoft with the intent of competing against the platform-independent capabilities of the Java environment.
2003	The complete draft of the human DNA sequence is completed by the Human Genome Project.

1

Before Computers

THE FIRST COMPUTER?

On Easter of 1900, a small group of Greek fishermen on their way home were pushed by a storm to Antikythera, a mostly uninhabited island north of Crete. Waiting out the storm, they did some sponge fishing in a cove on the island and found a large shipwreck. They notified the authorities, and archaeologists discovered that the ship, possibly Roman, probably sank between 100 and 40 BCE. In its hold were many statues of bronze and marble.

Archaeologists also found a curious corroded lump of metal that seemed to be the remains of a mechanical device. Many thought it to be an astrolabe, a device useful in navigation. In the latter half of the twentieth century, with the help of X-ray photography and painstaking research, Yale professor Derek de Solla Price (1922–1983) discovered the machine's true purpose. It was a mechanical computer for calculating lunar, solar, and stellar calendars. This was not a navigational device, but likely a prized part of the cargo. This find totally changed the historical view of when such complicated and potentially powerful devices could have been created. Historians previously thought that the level of sophistication shown by the Antikythera mechanism was not reached until the medieval European astronomical clocks of the fifteenth century.

Astrolabes were more common than the unique Antikythera device. The ancient Babylonians originally divided the circle into 360 degrees and developed the twelve signs of the zodiac, which the ancient Greeks then used to create the astrolabe, probably between 200 and 100 BCE. An astrolabe is a circular device that maps the heavens from a certain latitude. By rotating part of the device over the map, the user can determine current time, date, and latitude; determine positions of the sun and stars at any time of year; and calculate sunrise and sunset for any day of year. It can also be used to calculate distances and area related to the circle. It appears that the Antikythera device automated many of the relational calculations of an astrolabe through a series of gears and plates that showed the movement of the sun, stars, and moon.

At various times in the past, devices representing reality became very popular as both tools and objects of curiosity. Called automata, the mechanical recreation of reality, they probably existed even before Philon of Byzantium (circa 280–220 BCE) created a washstand that automatically dispensed a pumice stone and a set amount of water for washing. In the first century of the common era, Heron of Alexandria (circa 75 CE) created automatic theaters with mechanical figures acting out the play *Nauplius*. In the seventeenth and eighteenth centuries in Europe and America, many mechanical devices allowed for serious measurement while using their gearing to enact little scenarios, or even whole plays, with mechanical figures of people, animals, and natural phenomena like thunder, lightning, or waves. One of the most popular devices related to the astrolabe was the orrery—a device that accurately showed the movement of the planets around the sun. One of America's earliest scientists, David Rittenhouse (1732–1796) of Philadelphia, became famous in the eighteenth century for creating a beautiful and accurate orrery. Related to the orrery is the planetarium projection system that creates the image of the heavens on a domed ceiling—still used in planetariums the world over.

Another critical development in mechanical computing devices was tide predictors. Entering or exiting a harbor can range from annoying to dangerous depending on tides either running against your progress or taking your vessel dangerously close to submerged hazards. Creators of tables and charts did their best to prepare mariners with knowledge of tides by taking historical data and projecting them into the future. In the 1800s, Scottish physicist Lord Kelvin (1824–1907) created a formula for tides that he then modeled and refined with a machine that consisted of twelve pulleys (each pulley representing a coefficient of the equation) connected by a wire that was connected to a pen that drew the function (high and low tides) on a roll of paper. Each pulley was connected by a rod to a shaft that was turned to drive the machine. The gearing on the drive shaft could be changed to represent different locations on the earth. The U.S. Coast and Geodetic Survey created a similar machine with thirty-seven coefficients in 1911. That machine, 11 feet long and weighing nearly 2,500 pounds, was so successful that its accuracy (0.1 feet for any minute of a calculated year) was not matched until the mid-1960s by an IBM 7094 computer doing tens of millions of calculations for every year and location calculated.

The mechanical representation of mathematical formulas in the early twentieth century became most useful in what are called differential analyzers. Differential analyzers solved the problem of measuring the area under a curve. This problem can be so difficult to calculate that one technique used for a long period was to draw and cut out the curve on paper and then weigh the piece of paper—its weight was proportional to the area. A mechanical method developed by Lord Kelvin's brother James Thomson (1822–1892) was devised in the nineteenth century. His planimeter worked like a compass with measurements created by the friction of wheels as they rolled along the curve. The necessity of moving the device carefully without slippage made it only marginally accurate, though it was not replaced until the introduction of more accurate machining technology in the 1930s. Vannevar E. Bush (1890–1974) at the Massachusetts Institute of Technology created the first differential analyzer when he realized the speed it would give him in solving electrical power network problems, despite the long setup time of physically moving and rotating shafts for any particular differential equation. Bush directed the creation of three machines, the last completed right before World War II, and it was used throughout the war to calculate ballistic tables. The complexity of maintenance and programming the Bush differential analyzer, and the machine's relative slowness in calculating ballistics, prompted the U.S. Army to invest in a machine that is considered the first modern computer: the Electronic Numerical Integrator and Computer, or ENIAC.

Bush's differential analyzer inspired a number of copies and became instrumental in World War II. The copy at Manchester University in Great Britain was constructed out of a children's erector set. Another, at the Moore School of Engineering at the University of Pennsylvania, was more sophisticated than the original. A number were created in Russia, Germany, and Norway. Early in the war, components of the Norwegian machines were either stolen or destroyed by Norwegian resistance fighters so that they would not fall into German hands.

The Antikythera device, astrolabes, and other measurement devices show the importance of calculation and modeling in human history. The human drive toward improving calculation and modeling, along with numerous mechanical and electronic inventions, combined to eventually create the modern computer. Much of these early efforts concentrated on creating multipurpose calculating machines in order to make mathematics easier. The story of mathematics is an important precursor to the rise of the modern electronic computer.

MATHEMATICS

Along with language, mathematics has been a constant companion in human social evolution. Indeed, in many cultures, number systems were developed to a degree far greater than their use in basic needs. Often these numbering systems—such as those used by the ancient Mayans—were the province of the priestly class and used for complex religious ceremonies as well as having fun amongst themselves. The Greeks had two completely different systems: one used for numerical theory and another used for common purposes such as commerce. Esoteric mathematics has always been, and continues to be, known only by a few in society most highly trained in its use. However, over thousands of years, complex societies have required increasingly sophisticated mathematical skills at many levels of those societies. The trading of goods and services; the collection of monies for taxes; the building of structures such as pyramids, aqueducts, and skyscrapers; measuring property boundaries; creating and implementing instruments of war; navigating across land and water; and understanding time continue to be critical areas that require the widespread use of mathematics.

The origins of numeration, or counting, in the human species are lost in prehistory. The best evidence we have for counting before the development of writing is linguistic. Various language remnants still exist showing the frequent use of the numeric bases associated with 5, 10, or 20—the fingers (and toes) on one, two, or all four human appendages. Tribes as geographically separated as the Eskimos of Canada to various tribes in Indonesia have 20 as their numerical base. The Eskimos used the word "man" for each 20 units. Base 20 still exists in English in the word "score," perhaps most famous in Lincoln's Gettysburg Address opening, "Four score and seven years ago."

Many cultures have not moved beyond number systems tied to specific objects. For example, in the 1960s it was discovered that one language of the Indians in British Columbia had different systems for counting people, animals, canoes, smooth objects, long objects, and round objects. Evidence of this still exists in English as well: a "gaggle" of geese, for example, where the word "gaggle" is only used to describe a large number of geese and nothing else. One of the indications of a more advanced concept in numbers is having an abstract number system that is not tied to particular objects.

At some point, in many locations around the world, humans moved beyond the use of fingers and pebbles for counting purposes and began recording counted values. Today we have evidence of this through rock drawings, or notches on sticks that often refer to groups of things (bison or other hunted game, for example). This developed into groups of numbers represented by symbols. The Romans used M=1,000, D=500, C=100, L=50, X=10, V=5, and I=1. A later development allowed for the placement of the number to be used as aspect of value. As early as 200 BCE, the Babylonians had a zero place holder that they used to show a number as long as it had a nonzero digit in the ones (or units) place. Comparing to our decimal number system of today, this is similar to being able to have a number like 107, where the locations of 1 in the hundreds place, 0 in the tens place, and 7 in the ones place make sense, but not being able to use zeroes to pad out a number like 100. This inconsistency made the system less effective.

Our current decimal system, called the Arabic system, began in medieval Hindu India. The earliest physical evidence of the Hindi system is 876 CE, although it may have occurred much earlier. The Hindi system allowed for more complex math. The movement of this new system appears to have occurred fairly quickly—most likely due to extensive trade. Persians borrowed the system, and evidence exists that they transferred the system to Europe through Spain in 976. The use of the new system in Europe, however, was rare until the 1200s. Leonardo of Pisa (1175–1250), better known as Fibonacci, advocated the Arabic system in a book titled *Liber Abaci* (A Treatise on the Abacus) in 1202. Translations and variations of the book, *Arithmetic*, by the Arabic scholar Al-Khowarizmi (circa 780–850) finally succeeded in convincing great numbers of people of the usefulness of the system.

One reason the system was not widespread may have been the lack of Arabic translators for a number of centuries. The capture of Toledo in 1085 from the Moors may have helped solve the translator shortage. A number of merchants also resisted the new system. The city of Florence, Italy, prohibited the use of the new Arabic numbers in 1299 with the argument that they were too easily altered or forged. Roman numerals continued to be used extensively in Europe up through the seventeenth century, until the widespread adoption of the printing press proliferated and standardized the use of Arabic numerals and arithmetic. Roman numerals are still found in ceremonial forms today.

EARLY AIDS TO MATHEMATICS

Some of the arithmetic technology that evolved along with human society began quite early in human history and has continued on to modern times. Given the cost of paper during much of human history, and the difficulty in working with clay or wax tablets, it is not surprising that other techniques were invented than just written ones—especially for determining intermediate results in calculations.

One technique that many societies have used is tying various knots on cords to record numerical information. Biblical and Roman textual references indicate that those societies knew of the use of knots. Chinese records from as early as 2800 BCE indicate that knots were used until 300 BCE at least. The Peruvian Incas in the sixteenth century used possibly the most sophisticated knotted string system ever known. This system, known as a Quipu, consisted of a single string off which hung many knotted strings of different sizes and material. The Peruvians recorded everything from historical events to poems. German millers used a technique of knots to record flour sack contents until the beginning of the twentieth century.

Another technique relied on the use of tally sticks. Tally sticks are simply sticks or bone used as a surface for different carved markings. A tally stick known as the Ishango bone (actually notches on a bone), 8,000 years in age, has been found in Zaire. Other sticks or bones with markings have been found that are up to 30,000 years old, but they are less obviously tally sticks. The Chinese used tally sticks, and the remnants of those times remain in their language—the Chinese character for contract is "large tally stick." The British government used tally sticks from the thirteenth to the nineteenth centuries. All contracts were recorded by the Exchequer, and the Exchequer tally system included cutting various notches into a stick and then splitting the stick from end to end. The bank kept one side, called the "foil," and the individual kept the other side, called the "stock." This is where the term for owning part of a company as "stock" originated, and the term "tally up" as well. The centuries-old collection of sticks was finally deemed obsolete in 1826. In 1834, the British government began the process of burning the sticks in the stove of the House of Lords. Unfortunately, the sticks burned too hot and lit the paneling on fire, and soon the House of Lords and the House of Commons burned to the ground.

Knotted strings and tally sticks were probably used more for storing information than calculation, though they could be used in conjunction with other techniques to calculate. For example, the Chinese were probably the first to have a complete decimal system of numbers dating from as early as

1300 BCE. One technique the Chinese developed soon after the development of the decimal system was calculating rods. Rods could be made of wood, bamboo, bone, or ivory. Ivory rods were the most expensive and exclusive. These rods were combined in different ways to form the necessary numbers. The origins of the abacus include the use of these calculating rods. Later on, the rods were laid out on a board or cloth that was divided into squares. Each square would be a different digit of a number. By 800 CE, they added the zero to this system. The Chinese also had black rods to represent negative numbers and red rods to represent positive numbers. Red is a lucky color in Chinese; the term "in the red" would have the opposite meaning to what it means in the West. The Koreans and Japanese adopted the rods as well. The Japanese solved the problem of the rods rolling into the wrong square by flattening the rods.

In China, the rods were displaced by the modern form of the wire and bead abacus sometime around 1300 CE. The abacus seems to have its original origins in the Middle East, possibly modern-day Turkey or Armenia, during the Middle Ages. Tabular or board abaci existed in other regions much earlier. Because the Greeks did not write as much about common mathematics (logistic) as numerical theory (arithmetic), we do not have as much evidence or know as much about how the ancient Greeks calculated. However, the Salamis abacus found on the Greek island of Salamis, an approximately 2 by 5 foot marble table with numerical demarcations, is an excellent example of early abaci. References in Greek and Roman literature and paintings also place the first abacus at least as early as 400 BCE. Stones were used as counters on these boards at first. Later, in Medieval Europe, coinlike counters were used.

The French introduced "jetons" in the fifteenth century, commemorative sets of counting coins that were often given as gifts on New Year's Day. Jetons were exported to French colonies in North America until 1759 when the French suffered defeat at the hands of the British during what is known as the French and Indian War. Interestingly, the French took up the use of paper and pen using Hindu-Arabic numbers and abandoned the use of jetons soon after. By the 1812 invasion of Russia by Napoleon, French soldiers were bringing back commandeered Russian abaci as curiosities.

According to the Jesuit priest Joseph de Acosta (1540–1600), in Peru in 1590, some form of abacus existed in the Americas as well. Unfortunately, little is known about the Peruvian devices. In the right hands, the abacus could be manipulated quite quickly. In 1946, two individuals, Kiyoshi Matsuzake of Japan using the Japanese soroban abacus and Thomas Wood of the U.S. Army of Japanese Occupation using an electromechanical calculator, squared off for a contest in calculation speed and accuracy. Matsuzake won.

Another technique invented in many locations has been called finger calculation. Evidence suggests that finger techniques existed prior to 500 BCE in Greece. Because of the needs of trade across many cultures, an informal standardization of finger calculation likely occurred early in human cultural interaction. This allowed bargaining without the need to learn another language. A complete description of the system of finger calculation that was likely used from Europe to China was written by the English monk, the Venerable Bede (673–735), in 725 CE. The system used the left hand to represent from 1 to 99 and the right hand to represent from 100 to 9,900. Apparently, there were signs using the rest of the body that could represent up to 1,000,000, but they are not detailed as well. The system allowed for holding intermediate values on the hand as a temporary "register" while calculating in the head. Other finger systems existed as well. One system, used in Europe through the 1700s, could accommodate the well-educated people who still seldom knew the multiplication tables beyond five times five, and allowed multiplication of the digits from 6 to 10. This system was taught in some Russian schools into the 1940s.

The Greek invention of the astrolabe was likely originally designed for calculating time and location. However, it is also a device that can be used for calculation that uses the circle as a reference. Three later classes of devices also utilized these principles: the quadrant, the compass, and the sector. All three devices were used in one form or another into the twentieth century.

The Gunter's quadrant is an astrolabe consolidated onto one-quarter of a circle by Edmund Gunter (1581–1626) of Gresham College in London in the seventeenth century. Many quadrants were created with a number of scales on them, including trigonometric functions as lengths of lines. These were used in conjunction with trigonometric tables. Many also had squares and cubes and their roots on their back side. Large books were written describing all of the many functions that a quadrant could do.

The proportional compass was a set of dividers with the hinge between the two legs being an adjustable and scaled point—usually between 1 and 10. The scales allowed for obtaining squares or square roots geometrically—using lines, circles, or solids. For example, set the scale at 4 and measure a square of 1 inch a side on the small end of the compass, and you automatically obtain a square that will have four times the area on the other end. The origination of this device is lost, but documentation exists in the sixteenth century.

A sector was two scales hinged at one end. The Italian scientist Galileo Galilei (1564–1642) created one of the first around 1597. The initial use of the sector was by the military to calculate gun trajectories—not the last time an advanced calculation device was created for such a need. Generally, the device could measure an angle between 0 and 90 degrees for a gun's elevation. Zero degrees was left blank on the device, and so the term "point blank" came into existence. Later versions of the sector combined it with the compass and also had a curved interior to allow for measuring the size of the cannon ball. The number of scales also included cannon diameter or caliber, shot weight, amount of charge needed, and more. These devices also allowed for more generic calculations and could be used in conjunction with other sectors and various tables to do quite complex equations. The devices required a fair amount of skill in handling for advanced functions, leaving ample room for further improvement. In addition, precalculated tables were notoriously error prone, which also drove the need for more robust calculating tools. Tables were necessary, however, to simplify complex mathematics. Through a number of techniques, tables often turned difficult-to-calculate equations, including multiplication and division, into simpler-to-use equations of addition and subtraction.

JOHN NAPIER'S BONES

John Napier (1550–1617), born in Merchiston Castle, Scotland, is credited with creating at least two critical developments in the evolution of computation: logarithms and "Napier's bones." Born at the beginning of the Scottish Reformation, Napier spent much of his time managing the family estates and engaging in radical theological musings. He also found time for mathematics, and while some historians have speculated that logarithms were independently invented elsewhere, there is little evidence of such invention, or at least little evidence that it was communicated to Scotland.

Napier discovered that a series of numbers could be found that had a corresponding series where the numbers were what he termed "logarithms." His 1614 book, *Description of the Admirable Cannon of Logarithms*, known in its original Latin as the "Descriptio," described logarithms and included a series of tables with the logarithms of many numbers. This concept spread quickly and a number of more complete tables by other mathematicians followed. Henry Briggs (1561–1630) published a book of tables in 1624 that included the logs of numbers from 1 to 20,000 and from 90,000 to 100,000 to 14 decimal places. Johann Kepler (1571–1630), Edmund Wingate (1593–1656), Adrian Vlacq (1600–1667), and Edmund Gunter all added to and recalculated tables of logarithms.

To create the Descriptio, Napier invented another device to aid in calculation that he called the Rabdologia. The Rabdologia consisted of a set of plates that could be organized with respect to one another to give a multiplication product. The idea came from a much more ancient method of multiplication called gelosia where a matrix of multiplicands are created. Napier's Rabdologia became best known as "bones" because the best sets were created from ivory. Napier's bones only became known after his death in 1617 because Napier did not believe them worthy of publication. Once published in *Rabdologia*, however, many inventors furthered his work by creating more sophisticated sets of "bones." Gaspard Schott (died 1666), a Jesuit priest in Rome, published the ideas of Athanasius Kricher that extended Napier's bones with a creation he called the Organum Mathematicum in the 1660s. The Organum included bones for things as diverse as addition, subtraction, geometry, calendars, spheres, planetary movement, the construction of canals, the construction of military fortifications, and music. Schott also invented a box that put Napier's bones on cylinders, which ensured that the bones were correctly lined up and made calculation easier. The Englishman Samuel Morland (1625–1695) independently invented the same type of device in the mid-1660s. The technique of lining up Napier's bones in a mechanical way became a milestone in the creation of mechanical calculating machines. The most advanced version of Napier's bones were the Genaille-Lucas Rulers created by Henri Genaille and Edouard Lucas in 1885. These rulers eliminated the problem of the carry value between partial products.

Napier's bones also became instrumental in a device so potentially efficient and inexpensively produced that it later served engineers as their primary calculating device until the early 1970s: the slide rule. In 1620, Edmund Gunter created a scale of logarithms on stick of wood that could be used with a pair of dividers to easily add logarithms. The Englishman William Oughtred (1574–1660) simplified this invention by eliminating the need for dividers by having two scaled pieces of wood slide past each other, and thus became the father of the modern slide rule in 1622. Students of Oughtred created variations of these slide rules, one publishing before Oughtred did because Oughtred did not believe the work worthy of publication. A number of improvements on the slide rule continued over the next two centuries, but the device did not really replace the use of dividers and scales until the Scot James Watt (1736–1819) created an inexpensive and accurate slide rule in the late 1700s, and in 1850 when the nineteen-year-old Frenchman Amedee Mannheim (1831–1906) added the idea of the moving cursor over the slide. Mannheim's eventual appointment to professor at the École Polytechnique in Paris helped the slide rule become a mainstay device for mathematics.

MECHANICAL DEVICES

The eventual creation of the electronic digital computer drew heavily on the first mechanical automating of calculations. Mechanical calculators needed multiple parts: mechanisms to enter the number into the machine and select the correct motions for the correct function; a means to store the temporary values within the machine, including a possible carry; a method to display a result; and a method to reset the machine to zero. This all needed to occur with a minimum of human intervention. Unfortunately, early efforts at mechanical calculators suffered from less developed skills in machining and manufacturing. New gearing and techniques also needed to be invented to satisfy the tolerance requirements of the machines. The most skilled machinists, often found in the watch trade, were also not necessarily available for doing the work due to an initial lack of paying customers.

Although often attributed to Blaise Pascal (1623–1662), the first mechanical calculating machine probably belonged to Wilhelm Schickard (1592–1635), professor of math, astronomy, geography, Hebrew, and Oriental languages (as well as Protestant minister) in Tübingen, Germany. Schickard became fascinated with Kepler's descriptions of Napier's bones and created two machines that automated the multiplication process. While both machines are now lost, it became possible to reconstruct the machine from some of Schickard's notes in 1971. The machine worked. But from the reconstruction, a problem with early carry mechanisms became clear. The gearing used would potentially damage the machine if a carry needed to be propagated through the digits, for example like adding 1 to a number like 9,999.

The first calculating machine for which a copy still exists today came from Blaise Pascal. He worked out Euclid's geometric theorems on his own at age twelve, described complex conical geometry in a treatise when he was sixteen, and worked with Pierre de Fermat (1601–1665) to establish probability theory. In 1642, at the age of nineteen, Pascal invented a mechanical device for adding and subtracting, which he called the Pascaline, in order to assist his father in his job as tax collector of Normandy, France. To create the machine, Pascal had to train himself as a mechanic because local mechanics were not used to working to the fine precision needed for this machine. The machine used gears and wheels similar to the odometer in today's automobiles. The numbers appeared in small windows, and below those windows were dials similar to those on a rotary telephone. The operator used a stylus to turn the dials. The number windows actually displayed a choice of two values. If you examined the top number, you would have addition. Subtraction was done by observing the bottom number, which was the 9's complement. Pascal's mechanism did not allow for subtraction, since the gears could not run in reverse. The gearing did eliminate the carry problem obvious in Schickard's machine, however. Over the course of his life, Pascal made almost fifty different versions of the machine. All of them were fairly temperamental and required constant maintenance.

In 1694, Gottfried Wilhelm von Leibniz (1646–1716), a German rival mathematician of England's Isaac Newton (1642–1727), created a machine called the Leibniz wheel that worked similarly to Pascal's and also did multiplication. His first demonstration came in 1672 when he showed a wooden version that did not work very well to the Royal Society in London. A metal version was created in 1674 with the help of clockmaker M. Olivier. The machine introduced the concept of the stepped drum, a gear that has progressively deeper teeth, as a means to select the correct number. Unlike Pascal's machine, Leibniz's machine could move in reverse. Unfortunately, propagating a carry often required the user's intervention, so the machine signaled the user when a carry needed human intervention.

Samuel Morland realized his penchant for mechanical work when visiting the court of Queen Christina of Sweden in 1653, where he saw one of Pascal's adding machines. Morland worked on a version of his own and published his designs for three different machines in 1673. One machine was a mechanical adder. He elected not to create the complicated carry mechanisms of Pascal or the technique of Schickard. His machines were similar to Leibniz's in that they indicated to the operator when a carry propagation needed to occur. Because of this they were simple to operate and reliable, and could be made quite small; extant examples are 3 by 4 inches and only a quarter of an inch thick. Despite this convenience, few were sold. Another of Morland's machines automated Napier's bones. It allowed for the replacement of disks, a set of thirty in all that were essentially circular versions of the bones, that allowed for finding squares and cubes and their roots as well as multiplication and division.

The invention of a commercially successful automatic adding machine had to wait until the nineteenth century. In 1820, Charles Xavier Thomas de Colmar (1785–1870), serving in the French army, invented what he called the arithmometer. The arithmometer was a mechanically improved version of Leibniz's wheeled device. The size of a tabletop, the device could add, subtract, multiply, and divide. After leaving the army, de Colmar joined the new field of insurance. Using the calculation of insurance tables as an incentive, he continually improved the device, submitting it to various scientific competitions, and won the Legion of Honor. Some later devices had as much as a thirty-two-digit product register. Variations of this machine were used through World War I.

The simultaneous invention of a variable-toothed gear to replace the Leibniz stepped drum by both Frank S. Baldwin in the United States and Willgodt T. Odhner in Russia resulted in a considerable reduction in size and weight for calculating machines. Now calculating machines could sit on the corner of a desk. The Brunsviga company in the United States began manufacturing these machines in 1885 and sold 20,000 of them in the next three decades. Other equally successful companies also manufactured these types of machines. Mechanical improvements helped drive the market, including more functions and reduced size, and the comptometer, invented by Dorr E. Felt (1862–1930), improved the speed at which the keys could be pressed without the machine jamming. Until the advent of the digital calculator, these types of machines were the backbone of automated calculation on desktops in the Western world.

CHARLES BABBAGE

Charles Babbage (1791–1871) was born in England, showed an interest in both the internal workings of mechanical things and advanced mathematics as a child, and became a well-known mathematician by the 1820s. While a student at Cambridge University, Babbage and two friends translated, annotated, and added interesting examples to a French text on calculus that became the standard text for calculus instruction in Britain for most of the nineteenth century. After leaving Cambridge Babbage spent some time traveling, including a solo descent into the crater of Mount Vesuvius to spend the day taking numerous air pressure and temperature measurements while dodging venting hot air and lava. After moving back to London, he invented a cow catcher for the steam engines of the British railway system and an air conditioning system for his own London apartment. Babbage inherited a moderate fortune and did not have to worry about income for most of his life. A charming and busy member of London society as a younger man, he became ever more reclusive with age as he obsessed over his two greatest inventions: the Difference Engine and the Analytical Engine. He also became a permanent enemy of London street musicians when he argued to have them outlawed. For the rest of his life, he was hounded by street musicians who apparently took many opportunities to gather under his window in the middle of the night and play for him.

Babbage cemented his reputation in mathematics by producing a table of logarithms in 1826 that were the most complete and accurate yet published. Babbage focused not only on the correctness of the tables but also on their readability. He experimented with different type settings and the colors of type and paper. Some tables of logarithms at the time had more than 1,000 errors. The errata pages published to correct them often introduced even more errors. Despite his best efforts, Babbage's tables still ended up with approximately forty errors between his manuscript and the final printing. This did not satisfy Babbage, who felt that a machine might be the only way of removing the inevitability of human error.

After discussions with his friend, astronomer John Herschel (1792–1871), Babbage imagined reducing the errors present in some mathematical tables used by astronomers by creating a machine—possibly powered by steam—to do the calculations. He imagined a difference engine. The method of differences takes into account that values placed into the variable of a polynomial will create a constant difference in the value of the function. With simple functions that have a single term, the technique might require a single iteration through the equation. More complicated functions would require more iterations where the differences of the differences would have to be found. A difference engine that would cycle through the differences automatically had already been imagined by J. H. Muller, an engineer in the Hessian Army who had created a mechanical calculating machine. Muller's ideas were included in a book written by E. Lipstein in 1786. Muller failed to find funding for further development. In the 1820s, Babbage and many of his friends began to communicate with the Royal Society and the British government in support of his project to create a difference engine. As a demonstration, Babbage built a six-digit working model of his Difference Engine that could calculate to two levels of differences. At that time, the British government had a 200-year history of funding practical science and technology, and the government was interested in the Difference Engine to assist in creating actuarial, tide, navigation, engineering, logarithmic, and interest tables—all useful for the British Empire. The government advanced Babbage £1,500 in 1823, and Babbage agreed to personally fund the remainder of the project up to £5,000.

Babbage soon realized that the kind of machining necessary for the project would require expertise, and he found such an expert in Samuel Clement. Together, in pursuing the complex mechanisms needed for the engine, Babbage and Clement advanced the level of machine knowledge both in mechanisms and tooling in Britain. Many of the machinists apprenticed

Part of Charles Babbage's Difference Engine. Copyright Bettman/CORBIS.

in Clement's shops became quite successful. One, Joseph Whitworth (1803–1887), created the Whitworth standard for nuts and bolts. This is an early example of government investment in a technological project paying itself off in unexpected dividends. Unfortunately, the primary goal of creating the Difference Engine became complicated. Babbage lost his wife, daughter, and father, and his own health deteriorated, so he moved to Italy for a while to recover. Delays in getting funding from the government also created work stoppages. Babbage was forced to let go and then to hire and retrain staff during each stoppage. Finally, he built a shop on his own property to house Clement and staff, but Clement became concerned over the working arrangement and conditions, and their relationship deteriorated. Clement decided to leave, and British law supported him in taking all the tools with him.

During one of the stoppages, Babbage redesigned the Difference Engine. He applied for further funding to create a new version of the Difference Engine. The British government did not see the redesign in a positive light and cancelled any future funding in 1842. Babbage handed the plans and completed components of the first Difference Engine to the government. When completed, the machine would have had eighteen-digit numbers and have calculated to six differences. The machine would have automatically corrected round-off errors. It also had built-in safety features that would have stopped the machine from damage if it had any malfunctions. It would have printed the tables directly as well so as to avoid any errors in print setting. In 1991, the Science Museum in London took Babbage's detailed plans for his redesigned Difference Engine and built it: 10 feet long and 6 feet high, weighing 3 tons, and containing 4,000 parts. The Difference Engine worked with only minor changes in the design. The components for the first Difference Engine and a re-created second Difference Engine sit in Great Britain's South Kensington Science Museum today.

A number of difference engines were attempted by others in the following hundred years. Inspired by Babbage, the Swede George Scheutz (1785–1873) and his son Edvard created a small machine that could find three differences in 1843. In 1851, the Swedish government funded a complete machine with the stipulation that it be completed before the end of 1853. Their machine was completed by October of that year, and the Scheutzes improved its capability with an additional grant in 1854. In that year, the machine was examined by the Royal Society and praised by its members, including Babbage. After the machine won an award in Paris, it was purchased by the Dudley Observatory in Albany, New York, and was later purchased by Dorr E. Felt, inventor of the comptometer adding machine. Today the Scheutz difference machine is located at the Smithsonian Institution in Washington, D.C. A second duplicate machine was created by the Scheutzes in 1856. It was funded, ironically, by the British for £1,200 to assist the Astronomer Royal and the insurance industry. Today that machine is located at Great Britain's South Kensington Science Museum. Another Swede, Martin Wiberg (1826–1905), created a lighter and smaller version of the Scheutz machine in 1860. Others were also built, but by the

1930s, inventors realized that common desktop mechanical calculating machines could also be the basis for a difference engine.

While building the Difference Engine, Babbage began to imagine a different machine that he called the Analytical Engine. This new machine established the major logical components and techniques for the modern electronic computer, though it was never built. Today the proposed Analytical Engine is considered the first realizable design for a general-purpose computer. Its body was composed of three main components similar to today's computer's central processing unit (called the "control barrel" by Babbage), arithmetic processing unit (called the "mill" by Babbage), and memory (called the "store" by Babbage). The machine also had an input device for loading a program, using "Jacquard" punched cards and an output device that printed results. Programs could have controlled repetitive operations (called program iteration or program loops). An 1840 version of the design would have stood 15 feet high. The circular mill and control barrel structure would have had a radius of 6 feet, and the store would have extended out from this structure 10 to 20 feet. Card readers would have added to the size.

The control barrel (actually three different barrels) was similar to a music box cylinder. The studs on a music box pluck the right musical note. The studs on the control barrel would push various control rods. The rods connected to other parts of the machine, so that the barrel acted like the microcode of the machine. It created the processing sequence, directing the machine as to when to read from and write to store, read from the program, move register information to the mill, and any other necessary operations. Information from and to the registers passed along a rack of gears that simply transferred the settings of the individual registers to the mill. The mill would then perform the correct mathematical function, as directed by the program, on the numbers given to it.

The "store" had become obvious to Babbage as he constructed the Difference Engine. He realized that the registers that held each number to be manipulated were essentially identical. Instead of limiting the registers for particular functions, he created the store so that the registers could be used as a program directed—essentially creating the concept of random access memory. In addition, the memory was extendible. It was designed so that more registers could be added. The most detailed design by Babbage called for sixteen registers that could store either one 20-digit number or two 10-digit numbers. Other plans showed that Babbage may have envisioned fifty register columns of forty-digit numbers. There was also a special register that allowed for counting how many times a programming operation was performed—useful for doing program iteration.

Programs were input into the machine by the use of Jacquard punched-hole cards. Jacquard cards were rectangular, made of hard paper stock, and linked together with cloth along their long edges. The Frenchman Joseph-Marie Jacquard (1752–1834) essentially invented the first looping programmable machine in 1801. The machine was not used for calculation, however, but as a loom for weaving tapestries. Punched cards for weaving had been invented previously by Jacques de Vaucanson (1709–1782) in 1745, but his cards needed to be fed into the machine one at a time. What was later called the Jacquard loom used cards on a rolling drum. The cards would automatically feed into the machine, and as the cards passed a certain part of the machines, rods descended. The rods that could pass through the holes in the cards selected a thread to be woven into the tapestry. This meant that a deck of cards could be used repeatedly, creating the same pattern. Eventually, decks of over 24,000 cards that could create elaborate patterns were made. Napoleon awarded Jacquard for his work, and his business flourished. The professional weavers of the day, however, did not see the machine in quite the same glowing terms: Jacquard's life was threatened, and some machines were destroyed by vandals. Some versions of the machines are still in use today.

Babbage envisioned a number of Jacquard card readers attached to the Analytical Engine, with some readers potentially able to control the flow of others. The mechanism allowed reading the cards in sequence as well as out of sequence with the reading mechanism able to reverse the flow of cards as well. This feature meant that certain parts of the program might be repeated or skipped—thus establishing the three main types of advanced programming language statements: sequential, iterative, and conditional. A simple program on five cards might work as follows: store 2 in V1, store 3 in V2, read V1, read V2, multiply. The answer would print out as 6.

Babbage chose to think of the Analytical Engine as an academic exercise. None of the components designed was ever built by him, though the detailed parts of the final design seem to have been scrupulous in accommodating the machining techniques of his day. The parts would have required tolerances of one by five hundredth of an inch—possible, though expensive, in the mid-nineteenth century. Babbage continued to add to and modify the design for the remainder of his life, although only modifications to the main design occurred after 1840.

Some important details of the machine were written by other people excited by the project. While in Italy on a trip to Turin, Babbage described the machine to a number of interested engineers. A military engineer and future prime minister of Italy, Luigi F. Menabrea (1809–1896), wrote about the machine's operations in Italian. A friend of Babbage, the daughter of the poet Lord Byron, Lady Ada Augusta Lovelace (1815–1852), translated Menabrea's text into English and added considerably to the machine's operational instructions. Ada Lovelace has been called the "first programmer" for her efforts. In 1906, Babbage's son, Major Henry P. Babbage, had the R. W. Monro company create a version of the mill part of the machine to prove that it would work. It did, printing (using twenty-nine decimal places) the first twenty-five multiples of pi.

THE HUMAN COMPUTER

In the 1820s, while Babbage worked to create the Difference Engine, he traveled through Europe looking for manufacturing techniques that would suit his machine. He did not find much to add to the machine. However, in the process he became an expert in the latest manufacturing techniques throughout Europe. As a trained economist he saw these techniques in not only engineering terms but also economic terms. In 1832 he wrote the classic economic text, *Economy of Manufactures*, which along with the 1776 text *Wealth of Nations* by Adam Smith (1723–1790) were important sources of intellectual information as industrialists tried to understand the social forces that created the Industrial Revolution. The scientific management techniques of Frederick W. Taylor (1856–1915) in the late nineteenth and early twentieth centuries also relied on the insights of Babbage and Smith. Ironically, though Babbage wanted to remove people as much as possible from doing reliable calculations, his economic ideas were used to more efficiently organize the industrialization of information processing through the use of even greater numbers of people as human computers.

Babbage wanted to remove much of the human element from the creation of logarithmic, trigonometric, and countless other mathematical tables. In the nineteenth century and up through World War II, however, the term "computer" referred not to a machine, but to a person. Human computers used devices like the slide rule, abacus, and pen and paper; and later they used electromechanical devices like adding machines.

The first truly organizational approach on a massive scale to the creation of mathematical tables was probably that of Baron Gaspard de Prony (1755–1839) in 1790. In an effort to reform the tax and measurement systems after the French Revolution, Napoleon charged de Prony with creating these new tables, a huge undertaking. Inspired by Adam Smith, de Prony organized the effort along the lines of a factory with three sections. One section employed prominent mathematicians who determined the mathematical formulas to be used for calculating the tables. Another section used these formulas by helping organize the people who would do the calculations and collated the results. The third group consisted of up to eighty people who actually did the calculations. The calculations were made using the difference method, the same method that the Difference Engine used to break complex equations into equations with only addition and subtraction. This simplification allowed de Prony to hire less educated individuals for the third group. In fact, de Prony hired, in his own words "one of the most hated symbols of the ancient regime": the former hairdressers of the elaborate powdered wigs of the aristocracy.

Nothing on the scale of the French effort was known to exist in Babbage's England until the time that he wrote *Economy of Manufactures*, where he included a description of the Bankers' Clearing House of London. The increasing popularity of bank checks pushed the Clearing House, a secretive organization that Babbage managed to gain entry to through subterfuge, to prominence in London's financial circles. Almost £1 billion in 1830s currency were exchanged there every day. The sophisticated clerical organization allowed these exchanges between bankers. After the 1850s, this financial infrastructure model became more prevalent in Great Britain. In 1870, the Railway Clearing House had over 1,300 clerks. In 1875, the Central Telegraph Office had over 1,200 clerks, and four times that many in 1900. The rise of savings banks as a place to deposit discretionary income from the burgeoning industrial working class also increased the demand for clerical organizations skilled in processing large numbers of transactions with minimal errors. The Prudential Insurance Company took advantage of this type of processing in 1856 as well. For the first time, companies could offer insurance policies to working-class individuals because their processing organization made the administrative cost of individual policies less expensive.

HERMAN HOLLERITH

In the United States, no commercial entity approached the proportions of the European companies in terms of clerical needs. However, one governmental organization, the U.S. Census Bureau, did. An act of Congress in 1790 determined that a census of the population should occur every ten years so as to apportion members of the House of Representatives, one representative per 33,000 people. While the bureau employed fewer than thirty clerks in 1840 when the population was 17 million people, the bureau, along with the population, grew quickly after that. By 1880, the bureau employed almost 1,500 clerks to count a population close to three times that of forty years before. The census consisted of an army of census takers creating a huge amount of paper forms, which included information about an individual such as his or her sex, age, ethnic category, location, and so on. Back in Washington, D.C., the forms were collated on large tally sheets, a grid with rows and columns that corresponded to the various categories. Clerks created the final statistics after many passes through the forms. The whole process was done by hand with pen and paper.

In the 1880 census, the director of statistics, John S. Billings (1838–1913), suggested to a young staff member, Herman Hollerith (1860–1929), that the counting of the census should be automated by mechanical means. Hollerith had been a student at Columbia University's School of Mines. Between the 1880 and 1890 censuses, Hollerith worked at the U.S. Patent Office and as an instructor in mechanical engineering at the Massachusetts Institute of Technology. In his spare time, he constructed a prototype mechanical tabulating machine in anticipation of the 1890 census. By 1888, census officials concluded that another technique for the census had to be arrived at. The bureau had taken seven years to tally up the 1880 census, and feared that the 1890 census might not be completely tallied until after the 1900 census.

The new superintendent of the bureau, Robert P. Porter, sponsored a contest to replace the old system of pen and paper. Three individuals, including Hollerith, vied for the census contract. Two of the systems proposed were still human centered. They used different colored ink or cards and other techniques to make the process easier. Hollerith created a mechanical system inspired by his travels on railroads. When a conductor took a ticket from a patron, they punched holes in the tickets that corresponded to where the patron was sitting and physical characteristics of the patron like hair color and gender. It occurred to Hollerith that this system would work for the census. Hollerith proposed that clerks take the reports from census workers and create a paper card representing each individual in the United States by punching out the correct characteristics. The cards would then be tabulated automatically by a machine that would tally the characteristics by census district. The machine worked by passing metal rods through the card holes, a method similar to the Jacquard loom. In Hollerith's case, however, the rods dipped into small cups of mercury. The rods that made contact with the mercury passed an electric current through and thus incremented a electromechanical counter mounted on the front of the machine. The bureau contest was held in 1889 and consisted of reprocessing the St. Louis returns from the previous census. The three systems were fairly comparable in terms of speed when it came to taking the census reports and putting them on cards. After that phase, however, Hollerith's system was obviously so much faster and more flexible that he won the contract for the 1890 census.

Workers at the bureau created over 60 million cards for the census. While some of the newspapers were skeptical of the Hollerith machines, the bureau knew it had a winner. The bureau took only two and a half years for the census to be counted at a cost of $11.5 million dollars, $5 million less than the cost expected if the work was done by hand. Hollerith's machines were improved and used again for the 1900 census. In 1896 Hollerith created a company, the Tabulating Machine Company (TMC), and sold or leased the machines to other countries for their census taking. The machines also began to find their way into the private sector and were used to compile statistics for railroad freight, agriculture, and more. In 1911, TMC merged with another company and became the Calculating Tabulating Recording (CTR) Company. In 1910, an aggressive and successful

Two of the three machines that made up the Hollerith system. Copyright Hulton-Deutsch Collection/CORBIS.

salesman named Thomas J. Watson working for the National Cash Register Company (NCR) saw the potential of the CTR technology. By 1915, Watson led CTR. He focused the company on big leasing contracts, staying away from small office equipment sales. In 1924, he renamed the company International Business Machines Corporation (IBM).

IBM was one of the four business machine companies that dominated office equipment sales in the first half of the twentieth century in the United States. Of the four—IBM, NCR, Remington Rand, and Burroughs Adding Machines—IBM was the smallest. By 1968, though, IBM was bigger than the other three combined. In the late 1930s, IBM invested in the electromechanical Harvard Mark I, a realization of Babbage's dream of an Analytical Engine. Most of the calculations done in the twentieth century up through World War II, however, were not done by sophisticated general-purpose machines, for they did not exist yet. Equipment like IBM's tabulators and Burroughs' electromechanical adding machines fulfilled the increasing calculating needs of industrial nations, though these machines required multitudes of often unnamed human computers to operate in a systematic way.

2

The First Electronic Computers

THE ABC COMPUTER

When John Vincent Atanasoff (1903–1995) turned ten years old, his family moved into a new house at the phosphate mine where his father worked as an electrical engineer. For the first time, the young boy lived in a home with electrical lights. The family also bought their first automobile that year, at a time when most Americans could not afford automobiles. During that same year of wonders, his father also bought a new Diezgen slide rule, but found that he really didn't need such a sophisticated mathematical tool when he spent most of his time arranging for the repair of mining equipment and facilities. Atanasoff picked up the book of instructions and taught himself how to use the slide rule, performing additions, subtractions, multiplications, divisions, and then more sophisticated mathematical functions like logarithms and trigonometric functions. A book of his father's on college algebra helped him understand this new world of more advanced mathematics.

An avid reader, Atanasoff also found himself reading an old mathematics book of his mother's, where he was introduced to the idea of different bases for numbers. By convention, we today use base 10, with the digits 1 through 9 and 0, though pure mathematics does not require that people use base 10. The ancient Babylonians used base 60 (which is why we still have sixty seconds in every minute and sixty minutes in every hour), and the ancient Mayans used base 20. One of the bases that Atanasoff learned about was binary, or base 2, which uses only zeroes and ones to represent its numbers. The idea of number bases was at that time mostly a mathematical curiosity with little practical utility.

While still in high school, Atanasoff decided that he wanted to be a theoretical physicist, though when he attended the University of Florida, he found no classes in theoretical physics. The courses that electrical engineering offered were the most challenging, so he graduated in 1925 as an electrical engineer, like his father, and moved from the humid South to the dry plains of Iowa to start graduate studies in mathematics at Iowa State College, now known as Iowa State University. He earned a master's degree in mathematics from Iowa State, and a doctorate in physics from the University of Wisconsin in 1930, then returned to Iowa State College as an assistant professor of mathematics.

Atanasoff was frustrated by how long it took to calculate the results of a large number of calculations. Desktop mechanical calculators, manufactured by the companies of Monroe, Marchant, or others and powered by a hand crank, were used in these efforts, but mathematicians found them tedious to use and it might take weeks of work to solve a large set of equations. Desiring a machine that could solve partial differential equations, Atanasoff surveyed technology to see what might be available. An IBM tabulating machine in the statistics department used mechanical counters and intrigued Atanasoff, but he found that it really only added up categories of information on punched cards and could not solve equations. Atanasoff was the first to apply the word "analog" to machines that used mechanical counters. Digital computers store their numbers as distinct digits, with sharp boundaries between each number, whereas the numbers in analog computers have values that smoothly became other values.

Other analog mechanical computers included slide rules, the differential analyzer built by Vannevar E. Bush (1890–1974) at the Massachusetts Institute of Technology (MIT), machines that used Fourier analysis, and antiaircraft fire directors. The last were machines that calculated how far to lead the antiaircraft gun, based on the height and speed of the target, so that the shells would intersect with the aircraft. Atanasoff realized that he wanted to use a digital computer (though he did not coin the word "digital"). Digital computers that already existed included the Chinese abacus, some bookkeeping machines built by Burroughs, and the desktop mechanical calculators that he was already familiar with. Atanasoff thought of buying thirty or so Monroe mechanical calculators, arranging them in a line, and driving them simultaneously by a common shaft. The problem with this was that each calculation would have to be recorded by hand and inputted by hand on each machine, leading to a high risk of a mistake. Even a single mistake would render the calculation inaccurate.

Since what he wanted did not exist, Atanasoff decided that he would have to invent a computer, though he did not approach the project with enthusiasm. He was teaching both mathematics and physics, and was father to a family of three young children. He began to theoretically design such a computer and decided to use vacuum tubes, which were heavily used in radio technology but thought too unreliable by many experts for electronic applications. Vacuum tubes were electrical devices that could amplify electrical signals and act as switches. To make the computer more reliable, Atanasoff elected to use binary digits. Atanasoff was unusual in that he was familiar with base 2 when many other scientists and engineers were not. Atanasoff also wanted his computer to have an electronic memory as it made calculations (the same idea that Babbage called a "store") and coined the term "memory" in this context.

Frustrated with the lack of progress in his theorizing, Atanasoff took a long drive during the winter of 1937, traveling at excessive speeds, and eventually arrived in Illinois. After a couple of drinks at a roadhouse, the solutions to his problems became clear. He would use condensers (capacitors) for his memory, and to keep them from gradually losing the bit values put in the condensers, he would periodically pass electricity through the condensers to refresh them. He called this "jogging," and it is the same principle of refreshing used in modern computer memory chips today. He also decided to create logic circuits to perform addition and subtraction, instead of using enumeration as mechanical computers did. He drove home much slower, relieved to have broken through the mental barriers and have solved so many problems.

More months of theoretical introspection followed as Atanasoff expanded on the ideas that jelled during his roadhouse visit. In 1939 he received funding from the college to build a prototype and to hire an assistant. A brilliant young electrical engineering graduate student was recommended to him, and Atanasoff was fortunate to hire Clifford E. Berry (1918–1963). Working in a basement next to a student workshop, the two men carefully built each component and tested it thoroughly before moving on. They found vacuum tubes were expensive, used a lot of space, generated too much heat, consumed too much power, and broke often.

Atanasoff and Berry completed a working prototype before the end of 1939, and though it could only add and subtract binary numbers, the machine presaged the future. The computer was digital, used vacuum tubes, used binary numbers, used logic circuits, used refreshing memory, and had a rotating drum containing condensers to serve as memory. Rotating drum memory became popular for a couple of decades, although it is no longer used. The computer used a mechanical clock, driven by an electrical motor. The clock in a computer is like the metronome for a music student in that it keeps everything synchronized. Atanasoff wrote a manuscript in 1940 describing the theory of his computer, his plans for the future, and how the computer would solve large systems of linear algebraic equations. He used the manuscript to obtain more funding, and the Research Corporation awarded him a grant of $5,330. While the two men worked on creating a complete machine, a patent attorney was contacted. Iowa State College and Atanasoff agreed to share the patent, but uncertainty by the patent attorney over what documentation would be required for such a new device led to a delay in the patent application.

In December 1940, Atanasoff introduced himself to John W. Mauchly (1907–1980) at a meeting of the American Association for the Advancement of Science (AAAS) in Philadelphia because Mauchly had presented a paper on a harmonic analog analyzer that he had developed. The analog analyzer performed Fourier transforms. Atanasoff was excited to meet someone also interested in computing and invited Mauchly to come visit him in Iowa. Mauchly came during the summer of 1941 and stayed for five days, reading Atanasoff's research manuscript, examining the partially completed computer, and talking with Atanasoff and Berry about the invention. He also took notes on the manuscript, though Atanasoff would not let him have a copy of the manuscript because the patent application process was not completed.

So that human operators could work with their normal number system, Atanasoff and Berry added a device to the computer to convert to and from decimal (base 10) and binary (base 2) numbers. The computer could solve twenty-nine equations with twenty-nine unknowns and used punched cards, like IBM tabulating machines, to hold numbers beyond the capacity of their memory drum. Atanasoff and Berry completed their computer in early 1942, and it worked, though errors due to flaws in their punched-card stock occasionally cropped up.

The December 7, 1941, Japanese attack on Pearl Harbor changed everything for the two men. In June 1942, Berry moved to California to work at the Consolidated Engineering Corporation. In September 1942, Atanasoff moved to the Naval Ordnance Laboratory in Maryland, where he directed work on acoustics for the Navy for the rest of the war. Neither man ever returned to their work at Iowa State College. Sadly, Iowa State College failed to recognize the jewel in the basement, and the patent application was never actually submitted by the patent attorney. Since Atanasoff and Berry were gone, their computer was dismantled.

Last known photograph of the ABC computer before being dismantled by officials at Iowa State. Courtesy of the Charles Babbage Institute, University of Minnesota, Minneapolis.

Atanasoff occasionally met Mauchly during the war, and Mauchly told him that he was working on a computer based on completely different principles than Atanasoff's. We now know that was not true, since many of Atanasoff's ideas were incorporated into Mauchly's work. After the war, Atanasoff continued to work for the military in their Bureau of Ordnance, including an effort to build a computer for the bureau. The computer project was cancelled after a short time, and Atanasoff moved on to other projects. In 1949, he served as chief scientist for the U.S. Army Field Forces, and in 1952, he founded his own company, the Ordnance Engineering Corporation. In 1956, he sold the company to Aerojet General Corporation, and after a time as an executive with Aerojet, Atanasoff retired in 1961.

Mauchly and his colleague, J. Presper Eckert (1919–1995), filed for a patent on their ENIAC computer in 1947, and for many years historians considered the ENIAC to be the first electronic digital computer. This patent was later owned by the Sperry Rand Corporation, and eventually lawsuits began when Sperry Rand asked for royalties from other computer manufacturers. Lawyers contacted Atanasoff and asked for his help, prompting Atanasoff to examine the ENIAC patent. He was surprised to find many of his own ideas in the patent and participated in an epic legal battle to overturn the Mauchly and Eckert patent. Berry did not participate because he had apparently committed suicide in 1963 for unknown reasons. To honor Berry's contributions to their joint effort, Atanasoff started to refer to their computer as the Atanasoff Berry Computer or ABC computer. In a 1973 federal court decision, the Mauchly and Eckert patent was set aside and the ABC computer declared the first electronic digital computer. The judge handing down this decision did not know of efforts in Germany with the Zuse machines and in Britain with the Colossi that could also lay claim to being the first electronic digital computers.

Though Atanasoff received no royalties from inventing the computer, he was showered with awards after the 1970s. The communist nation of Bulgaria, proud of a man who had an obvious Bulgarian name and was the son of a immigrant from Bulgaria, awarded Atanasoff their highest scientific honor. Among other awards and honorary degrees that graced Atanasoff's later years were the Pioneer Medal from the Institute of Electrical and Electronics Engineers (IEEE) Computer Society in 1984, an IEEE Electrical Engineering Milestone in 1990, and a Medal of Technology in 1990 from the U.S. Department of Commerce.

CODEBREAKING WITH BOMBES AND COLOSSI

Sailors have always had reason to fear the sea, and stories of shipwrecks and ships lost at sea, victims of rocks or harsh weather, are common fare. During World War II, Allied sailors in the Atlantic experienced an additional fear, as thousands of them lost their lives to torpedoes from German U-boats. Unknown to these men, as they waited in the darkness and wondered if they would live, some of the first computers were helping them to survive. These computers remained unknown to historians for decades after the end of the war.

World War II was a war of science and technology as much as it was a struggle between fighting men. Part of that struggle involved codebreaking. Competent military commanders have always tried to keep their plans secret from their enemies, which led to the rise of codes to conceal the contents of written messages. With the coming of radio communications in the twentieth century, codes were applied to radio traffic to conceal their content from enemy eavesdroppers. Codes became so sophisticated by World War II that mechanical help was required to encode and decode messages.

The Enigma encoding machine was patented in 1919 by a German company for commercial use, concealing messages from possible business competitors. In 1926, between the two world wars, the German Navy adopted the Enigma machine to encode their radio traffic. Other branches of the Germany military and other German government departments followed suit in the next decade. The Dutch military also purchased Enigma machines and began to use them in 1931. In 1943, the Germans shipped 500 Enigma machines to the Japanese for use in German-Japanese communications. Limited numbers of Enigma machines were also used by Italy and other German allies.

The electromechanical Enigma looked like a portable typewriter in a small wood box. A lamp board above the keyboard contained the twenty-six letters of the alphabet. A set of three rotors above the lamp board could each be rotated to twenty-six different positions. The operator set the rotors to a daily prearranged setting, typed in a message, which sent voltage through the machine letter by letter, and the encoded or decoded letters appeared on the lamp board. After encoding a message and copying the resulting letters down, the operator then sent the message via telegraph or via radio using Morse code. The message was transmitted in the clear but read like gibberish. Because the Enigma encryption scheme was symmetric, the receiver of an encoded message only needed to have the correct rotor settings in order to decode the message. As an operator typed the encrypted gibberish into an Enigma machine, the lamps on the lamp board lit up with each of the decoded letters in correct order.

The German military added a plugboard to their Enigma machines that allowed up to six pairs of letters to be interchanged. The plugboard, combined with the three rotors, resulted in many millions of possible combinations. Certain that no one could break messages sent by their machines, the Germans allowed the continued sale of commercial Enigma machines. The Poles detected that the Germans were using this new machine and bought one themselves, added the military plugboard to it, and tried to figure out a way to break the daily code settings. They failed until the mathematician Marian Rejewski (1905–1980) began working on it and developed a decryption technique that often worked to decode a set of messages. The Poles also built an electromechanical machine to help them in their work. The ticking sound of the machine prompted them to call it a bomba, the Polish word for bomb. When Adolf Hitler decided to invade Poland, his intentions became obvious to the Poles from decoded radio intercepts. Six weeks before the war began, the Poles called a meeting with their allies and revealed to the British and French the extent of their success. The Poles gave the British a copy of the Enigma machine, the plans for their bomba, and a copy of the statistics that they had gathered that allowed them to more easily break new Enigma traffic. Remarkably, even after Poland and France fell to the German armies within a year of this important meeting, the Polish breakthrough remained a secret.

The British realized the jewel that had been handed to them and created an organization to exploit it. They built large radio receivers to pick up radio messages bouncing off the atmosphere from deep inside Nazi-occupied Europe. These messages were transcribed and carried to a government-owned estate outside of London called Bletchley Park. The British recruited the best and brightest to work in spartan conditions at Bletchley Park to decode the Enigma traffic. Women from the auxiliaries of the British armed forces formed the backbone of the effort, hunched over radio receivers and typewriters, engaged in the detailed work of indexing thousands of radio intercepts a day and keeping the flow of paper going.

The Germans usually changed their Enigma rotor and plugboard settings every day, using printed code books so that distant military units knew the settings for each day. They also made the rotors removable, so that a set of five (eight on naval models) rotors were supplied with each machine, with only three being used each day. Different branches of the German military used different codes, so that the navy, weather service, different commands of the army, and so on would have different codes for a given day. This meant that the wizards at Bletchley Park had to break the code for a given day and for a given service, leading to a never-ending effort. Mistakes by German operators and standard formats for certain types of messages helped the British codebreakers. At times the British even planted information, hoping to create a situation where the codebreakers might already know the content of an encrypted message.

The brilliant British mathematician Alan Turing (1912–1954) served as a leading codebreaker at Bletchley Park. Turing's skills in mathematics were recognized at an early age, though he was not a good student at the boarding schools that he grew up in. After a couple of failures to gain admission to college, he was accepted at King's College at Cambridge University and graduated with a master's degree in mathematics in 1934. His 1936 paper, "On Computable Numbers," contained an argument about what kinds of problems are computable and proposed a theoretical computer that became known as the Turing Machine. After a couple of years studying in the United States at Princeton University, Turing returned to Cambridge and became involved in theorizing efforts on how to build a computer. His colleagues and he decided to use binary numbers and boolean algebra. A reflection of an occurrence common in the history of technology, Turing, Atanosoff, and other computer pioneers all stumbled onto binary as the solution to making electronics of their computers simpler.

When World War II broke out, Turing was recruited into Bletchley Park. Turing and the British were completely unaware of the obscure efforts of Atanasoff when Turing and other codebreakers redesigned the bomba, which the British called a bombe. The first bombe was built in 1940 at the British Tabulating Machinery factory in Letchworth. The British bombes were essentially electromechanical reproductions of twelve enigma machines each, emulating the rotor settings. These bombes did not break the code, but excluded possibilities, leaving the remaining possibilities to be broken by hand. The noisy bombes broke frequently and required almost constant repair. The British eventually shared their secrets with the United States, and the Americans built their own versions of the bombes. Faster bombes that emulated up to thirty-six enigma machines and that weighed over a ton were built by both the British and the Americans in 1943. Over 100 bombes in total were built during the war.

Besides the Enigma machines, the German Army also started to use Lorenz SZ42 cipher machines during the war, especially for high-level communications between Berlin and distant armies. These machines encrypted their teleprinter traffic through an encryption system invented by an American, Gilbert Vernam, during World War I. The Lorenz machine

Photograph of a bombe. Courtesy of the National Security Agency.

was superior to the Enigma machine in that it both encrypted and transmitted its teleprinter traffic, as well as automatically received and decrypted the messages at the receiving end. This system relied on a set of randomly created characters that were interspersed with the clear text before encryption. At the receiving end, the same set of randomly created characters were used to decrypt the message. If a truly random set of characters is created, then copied for use by the sender and receiver, the system is theoretically unbreakable. The Lorenz engineers realized how difficult distributing such random sets were, so they built into the machine the ability to create a pseudorandom character set. These automatic pseudorandom characters sets weakened the strength of the cipher. Whereas Enigma machines were used mainly for tactical purposes, the Lorenz machines were used to transmit longer communiques of strategic value, like order of battle information, supply reports, and military planning discussions.

The British detected radio traffic from the Lorenz machines in 1940 and codenamed the unknown messages "Fish." Codebreakers at Bletchley Park figured out how to break the code, but the process took so long that the decrypted messages might be weeks old and the intelligence grown stale. The codebreakers built a machine called the Heath Robinson (named after a cartoonist known for his drawings of fanciful machines) that read in two paper tapes, slowly searching for a match. Tommy Flowers (1905–1998), a mechanical engineer and telephone exchange expert at the Post Office Research Labs at Dollis Hill, took the concept from there, building the Colossus. Users programmed the digital Colossus via its plugboard and switches. The Colossus was also an example of an early computing effort to manipulate symbols in the form of letters rather than just serve as number crunchers.

The original Colossus machine used 1,500 thermionic valves (vacuum tubes). A paper tape with an encrypted message on it was fed into the machine at 5,000 characters per second. The British built a total of ten Colossi during the war, eight of which were the more advanced Mark II version with 2,400 vacuum tubes. The Mark II machines ran five processing units in parallel, reaching an effective total speed of 25,000 characters per second. The Colossi decrypted a total of 63 million German characters before the end of the war. By 1943, Bletchley Park was regularly reading Fish traffic after a delay of only a few days. What is remarkable about the effort to break the Lorenz machines is that the British succeeded without ever capturing or learning any details about the actual machines.

The British and Americans used the term "Ultra" to describe decrypts that came from Bletchley Park. Keeping Ultra secret was considered as important as keeping the Manhattan Project to build the atomic bomb secret, and Ultra intercepts significantly helped the Allies win the war against Germany. After World War II, the secrets of Bletchley Park were not released to the public. Eight of the Colossi were immediately destroyed, and the last two were destroyed in about 1960 and their blueprints burned. In the 1970s, details about Bletchley Park slowly became known. As part of efforts to reclaim this lost history, a new Colossus was built at a museum in England, relying on memories of surviving engineers, what few pictures and diagrams were not destroyed, and information that the Americans had retained about the machines.

Turing's contribution to the success of Bletchley Park was also kept a secret. After the war, Turing continued to work on computers and spent time working on a Automatic Computing Engine at the National Physical Laboratory before moving on to serve as deputy director of the Royal Society Computing Laboratory at Manchester University. In 1950, Turing published a famous paper on artificial intelligence that proposed a test for determining if a computer was intelligent. After being convicted of a crime related to his homosexual behavior and serving his sentence, Turing committed suicide by eating an apple laced with cyanide.

THE ZUSE COMPUTERS

Konrad Zuse (1910–1995) was born in Berlin and had dreams of designing great cities or moon rockets. He attended the Technische Hochschule of the University of Berlin and became a civil engineer in 1935. While in school, Zuse conceived of a machine to automatically solve systems of linear algebraic equations, and in 1936 he began to design a mechanical computer that he later called the Z1. He chose to use binary, coming to this idea a year before Atanasoff in Iowa. During the day, Zuse worked as an aircraft engineer conducting stress analysis for the Henschel aircraft company; at night and on weekends he worked on his computer, funding the effort out of his own salary. A friend, Helmut Schreyer, helped Zuse in his work.

Zuse chose to leave Henschel in order to devote more time to his computer project, and in 1938 finished the Z1 in the living room of his parents' home. Instructions were fed into the machine by using old movie film as punched tape. Memory was maintained via pins in small slots cut into sheet metal. The machine used a floating point format to represent complex numbers, a significant innovation compared to the effort by Atanasoff. The arithmetic unit of the machine could only work for a few minutes before errors cropped up. The patent claim that Zuse filed with the U.S. Patent Office was rejected because of insufficient detail.

Zuse and Schreyer did not give up, but began to work on their next computer, the Z2. This computer was electromechanical, using secondhand relays. They wanted to use vacuum tubes, but the number they required were beyond their financial means. The coming of World War II resulted in Zuse being drafted into the army, and a year passed before he was discharged to return to his old job with Henschel. Zuse completed the Z2 on his own and demonstrated it to the Germany Aeronautical Research Institute. The German government agreed to fund his effort to build a Z3, but did not give him sufficient support to do more than continue to work at his home.

The Z3 was finished before the end of 1941 and contained 2,600 relays, but still used mechanical memory. The Z3 was faster in multiplying numbers than the electromechanical Harvard Mark I, but much slower than the all-electronic ENIAC. An air raid destroyed the Z3 in 1944.

Given more support, Zuse founded a company, Zuse Apparatebau, in 1942 and started to build the Z4. In the meantime his company built several special-purpose calculators for Henschel to calculate wing and rudder surfaces for aircraft and flying bombs. The coming end of the war caused Zuse to flee to the small town of Hinterstein in Bavaria, where he hid the dismantled Z4 computer in the basement of a farmhouse. The Z4 was eventually retrieved and completed, being set up for use at the Federal Polytechnical Institute in Zurich, Switzerland, in 1950 and then being moved five years later to the French Aerodynamic Research Institution, where it was used until 1960.

During the chaos after the end of World War II, Zuse found the time and mental focus to create one of the first computer programming languages, which he called Plankalkul. Historians have been impressed by a language that used variables, conditional and looping statements, and procedures. Zuse also formed a company after the war, Zuse Kommandit Gesellschaft (Zuse KG), which continued to make mechanical computers for the European market. In 1958, the Z22 became the first Zuse computer based on vacuum tubes. Zuse KG eventually became part of the large German firm of Siemens. While Zuse's efforts were impressive, from a historical point of view, his machines did not influence the development of later computers. In a sense, his efforts were a historical dead end, a result of being isolated from the dynamic technological innovation going on in the United States and Britain.

THE HARVARD MARK I

Another computer pioneer, Howard Hathaway Aiken (1900–1973), was born in Hoboken, New Jersey, but grew up in Indianapolis. In high school, he began working for the Indianapolis Light and Heat Company. He continued to work in the electrical utility industry while in college at the University of Wisconsin–Madison. Graduating as an electrical engineer in 1923, he remained in the same field of work for another nine years, when he decided that he had picked the wrong career. He wanted to be a physicist, so he enrolled for a year at the University of Chicago to study mathematics and physics before moving on to Harvard University. He obtained his master's degree in 1937 and a doctorate in 1939.

Like other computer pioneers, Aiken wanted to solve large systems of equations, leading him to think about building a computer. He began to work on a calculating machine when a technician took him into an attic at Harvard and showed him a piece of the Babbage calculating engine that had been donated to Harvard by Babbage's son in 1886. Aiken was fascinated and studied the work of Charles Babbage, coming to view himself as Babbage's successor. Aiken approached Thomas J. Watson Sr. (1874–1956), the president of IBM, for funding. Watson generously supported Aiken's effort, and IBM engineers did most of the design work as well as actually build the machine. Unlike the mechanical differential analyzer built by Vannevar E. Bush (1890–1974) at MIT, Aiken's proposed electromechanical machine was designed to perform all kinds of mathematical operations, not just differential equations.

After World War II started, Aiken served in the Navy before being asked to return to Harvard and direct the U.S. Navy Computing Project. After spending half a million dollars, the Mark I was completed in 1943. IBM had paid for two-thirds of the cost, while the Navy picked up the rest of the total cost. The long narrow machine stretched 51 feet from side to side, stood 8 feet high, and was only 2 feet deep. Weighing 5 tons, the Mark I included 3 million wire connections. The heart of the machine was seventy-two IBM mechanical rotating registers. Unlike the British bombes, the machine was as quiet as a few typewriters. Programming the machines required adjusting 1,400 switches, using a paper tape to feed in instructions, and using punched-card readers to input data.

Grace Hopper (1906–1992) worked closely with Aiken on the Mark I computer and later became an important figure in the development of programming languages. Hopper, born Grace Brewster Murray in New York City, attended private schools for girls, then the women-only Vassar College in Poughkeepsie, New York, before attending Yale University in 1928. In

1934, she graduated with a doctorate in mathematics and mathematical physics from Yale. Social attitudes against women in the workplace limited opportunities for women at that time, and she turned to teaching at Vassar. She married in 1930, taking the last name of Hopper, though the childless marriage ended in divorce in 1945.

In 1943, Hopper joined the Navy and was assigned to the Navy's computer project at Harvard a year later. Aiken assigned her to read the writings of Babbage and to write the manual for the Mark I, which led to her programming the Mark I. Aiken and Hopper published joint scholarly articles on their efforts, establishing her reputation as a programmer. Aiken went on to build the Mark II for the U.S. Navy, based entirely on electromagnetic relays, and several more Mark computers in the 1950s, each using ever more advanced technology. While Aiken's efforts led to useful computers, and the Harvard Mark I was one of the most impressive electromechanical computer ever built, the modern computer, with stored programs and entirely based on electronics, traces its lineage from the ENIAC computer.

THE ENIAC

Mauchly and Eckert formed a team that managed to bring the electronic computer out of government laboratories into the commercial world of data processing, where they launched a revolution. John W. Mauchly was born in Cincinnati, Ohio, and grew up near Washington, D.C., where his father was a physicist at the Carnegie Institute. He enrolled in Johns Hopkins University in Baltimore, Maryland, in 1925 and after two years as an undergraduate he applied to enroll directly in a doctoral program in physics. He earned his doctorate in 1932, emphasizing the study of molecular spectroscopy, and joined the faculty of Ursinus College in Collegeville, Pennsylvania. Like most scientists, he was frustrated by how long it took to solve large systems of equations, especially since his research focus had turned to studying weather systems. He hired a group of graduate students in mathematics to use mechanical calculators to solve the large number of equations he needed in order to understand the statistics behind the effects of a solar flare on the weather. That tedious exercise prompted him to build a machine to solve Fourier transforms and led him to present a paper on it at the AAAS Philadelphia meeting, where he met John A. Atanasoff. After he returned from examining Atanasoff's work in Iowa, Mauchly took an advanced course in electronics at the Moore School of Electrical Engineering at the University of Pennsylvania and was asked to remain on the faculty of the school. Mauchly met Eckert soon after arriving at the Moore

School. A native of Philadelphia, J. Presper Eckert completed his bachelor's degree at the Moore School in 1941 and had remained at the school for his graduate studies.

The Moore School signed a research contract with the Ballistics Research Laboratory (BRL) of the U.S. Army, and in an August 1942 memorandum Mauchly proposed that the school build a high-speed calculator out of vacuum tube technology for the war effort. In 1943, the army granted funds to build the Electronic Numerical Integrator and Computer (ENIAC) to create artillery ballistic tables. Eckert served as chief engineer of a team of fifty engineers and technical staff on the ENIAC project.

Completed in 1945, the ENIAC consisted of 49-foot-high cabinets, almost 18,000 vacuum tubes, and many miles of wiring, and weighed 30 tons. In order to minimize the high failure rate of vacuum tubes, Eckert ran the tubes at a lower voltage than they were designed to handle. A plugboard was used to program the computer. A ballistic trajectory calculation that took a human mathematician twenty hours to solve was completed by the ENIAC in 30 seconds. Oddly enough, the ENIAC used decimal numbers instead of binary numbers.

The ENIAC computer. Courtesy of the Charles Babbage Institute, University of Minnesota, Minneapolis.

The American ENIAC machine was completed two years to the month after the first British Colossus, which was an equivalent machine in many ways, though the ENIAC engineers had no idea that the Colossi existed. If the ABC computer was a rowboat, the ENIAC was a three-masted sailing ship ready to carry cargo, though many of the key original innovations came from the work of Atanasoff and Berry. While building the ENIAC, Mauchly and Eckert developed the idea of the stored program for their next computer project, where data and program code resided together in memory. This concept allowed computers to be programmed dynamically so that the actual electronics or plugboards did not have to be changed with every program.

The noted mathematician John von Neumann (1903–1957) became involved with the ENIAC project in 1944 after a chance encounter with an army liaison officer at a railroad station. Johann (later anglicized to John) von Neumann was born in Budapest, Hungary, to a Jewish family, where his father was a banker. His family recognized his extraordinary intelligence as a child and hired a private tutor to supplement his education. When he received his doctorate in mathematics from the University of Budapest in 1926, von Neumann was only twenty-two years old and already publishing mathematical articles. In 1930, von Neumann moved to Princeton University in the United States, where he began as a visiting lecturer, becoming a full professor and an original member of Princeton's Institute for Advanced Study only three years later. His prowess in mathematics and numerous original contributions made him a leading theorist in game theory and set theory. During World War II, von Neumann worked on the Manhattan Project to build the atomic bomb and also lent his wide expertise as a consultant on other defense projects.

After becoming involved in the ENIAC project, von Neumann expanded on the concept of stored programs and laid the theoretical foundations of all modern computers in a 1945 report and through later work. His ideas came to be known as the "von Neumann Architecture," and von Neumann is often called the "father of computers." The center of the architecture is the fetch-decode-execute repeating cycle where instructions are fetched from memory and then decoded and executed in a processor. The results of the executed instruction change data that are also in memory. After the war, von Neumann went back to Princeton and persuaded the Institute for Advanced Study to build their own pioneering digital computer, the IAS (derived from the initials of the institute), which he designed.

Along with honorary doctoral degrees, von Neumann was elected to the National Academy of Sciences in 1937 and received several national honors for his defense work. He died of cancer in 1957, perhaps contracted during his work on the Manhattan Project. Von Neumann was a gregarious man with sophisticated tastes, a command of four languages, a prodigious memory, and an amazing ability to perform calculations in his head.

Eckert and Mauchly deserve equal credit with von Neumann for their innovations, though von Neumann's elaboration of their initial ideas and his considerable prestige lent credibility to the budding movement to build electronic digital computers. The Electronic Discrete Variable Arithmetic Computer (EDVAC) was designed by the Moore School as the successor to the ENIAC and to be the first stored program computer, using binary numbers instead of decimal numbers, but the departure of Mauchly and Eckert in 1946 to form their own commercial venture delayed the completion of the EDVAC until 1952.

THE MANCHESTER MARK I AND THE EDSAC

The Moore School decided to hold an eight-week summer school on the "Theory and Techniques for the Design of Electronic Digital Computers" during the hot months of 1946, instructing invited scientists and engineers in the new art of computers. This onetime event, later known as the Moore School Lectures, effectively spread the knowledge created at the Moore School and prompted other digital computer efforts. Multiple mathematicians and engineers from Britain visited the Moore School or attended the lectures. Some of these British scientists and engineers had worked on the bombes and Colossi of Bletchley Park, and so had additional knowledge that they were required to keep secret. The others had usually worked on the war effort in some capacity. F. C. Williams (1911–1977) visited the Moore School that summer and noted the work on creating memory storage via a mercury acoustic delay lines or by using cathode ray tubes (CRTs). Williams returned to the Telecommunications Research Establishment (TRE), where he had worked on radar and Identification Friend or Foe (IFF) aircraft systems during the war. Working with his colleague, Tom Kilburn (1921–2001), Williams developed a way to store binary bits inside CRTs, which became known as Williams' tubes.

Cathode ray tubes work by shooting an electron beam at a screen covered with phosphor dots. The electrons in the beam react with the phosphor atoms to release photons as light. This effect persists for a fraction of a second and will disappear unless the electron beam refreshes the phosphor dot. The Williams' tubes took advantage of this delay to store the values of bits on the surface of the CRT. By changing the intensity of the electron beam from a writing mode to a sensing mode, the beam could sense whether a bit was displayed on the screen. Because the effect of the electron beam on the phosphor atoms rapidly deteriorated, memory had to be regularly refreshed by sensing the bit values and then rewriting them.

Williams and Kilburn both moved to Manchester University, where a group of mathematicians and engineers who had worked on radar research or at Bletchley Park during the war came together to work at the Royal Society Computer Laboratory. Alan Turing also joined them for a time. In 1948, the Manchester team succeeded in running a simple stored program on their prototype machine. Their machine had only three CRTs, one of which stored thirty-two words, each thirty-two bits long, with the other two CRTs being used as an accumulator and to hold the instruction under execution. After completing their prototype, later called the Manchester Mark I, based on Williams' tube electrostatic memory units and the use of a rotating drum to magnetically store data, the Manchester team contracted with an outside firm to build the first production model. The production model was also called the Mark I, and was built by a local electronic firm, Ferranti. The first Ferranti Mark I, delivered to Manchester University in spring of 1951, actually beat the more famous UNIVAC as the first commercially produced computer to be delivered to a customer. The Ferranti Mark I contained 4,000 vacuum tubes, 2,500 capacitors, and 12,000 resisters. A team at Manchester University also continued to develop more sophisticated computers, laying the foundation for a commercial computer industry in England.

The British mathematician Maurice V. Wilkes (1913–) also attended the Moore School Lectures, and on the voyage home to Cambridge University began to design his own computer. His goal was simply to create a stored-program computer, though Wilkes correctly perceived that the main technical problem was how to store memory. The Cambridge team led by Wilkes developed a way to temporary store an electrical impulse in a tank of mercury, the mercury-delay line.

The Electronic Delay Storage Automatic Calculator (EDSAC) performed its first calculation in 1949, running the first program written for a fully functional stored-program computer. The EDSAC beat the EDVAC and UNIVAC projects, making its mark in computer history as the first functional stored-program computer. The Cambridge team rapidly innovated in programming, developing a symbolic notation system and even the idea of subroutines, which are segments of reusable code. Wilkes and two colleagues wrote the first book on programming, *The Preparation of Programs for an Electronic Digital Computer*, in 1951. This textbook strongly influenced other projects, both in Britain and back in the United States.

THE UNIVAC

Mauchly and Eckert managed to persuade the Census Bureau in 1946 to contract for a computer called the UNIVAC (UNIVersal Automatic Computer). The Bureau contracted to spend $300,000, though Mauchly and Eckert thought that developing the machine would cost $400,000. They hoped to recover their financial loss by selling more computers. Hiring engineers and technicians, they worked long hours as financial difficulties beset their Philadelphia-based company. A contract with Northrop Aircraft Corporation to build a small guidance computer called the BINAC (BINary Automatic Computer), for a guided missile, helped keep the company afloat. In 1948, the A. C. Nielsen market research firm and the Prudential Insurance company also contracted to buy UNIVACs, though for substantially less than the U.S. Census Bureau paid. Desperate for money to finance their research and development activities, Mauchly and Eckert sold 40 percent of their corporation to the American Totalisator Company of Baltimore for $500,000 and additional loans to save the company.

By 1949, the Eckert-Mauchly Computer Company was doing well with 134 employees and six orders for their UNIVAC, still under development. The death of the founder-president of American Totalisator brought

The UNIVAC computer. Courtesy of the Charles Babbage Institute, University of Minnesota, Minneapolis.

financial calamity as the subsequent management of American Totalisator called in their loans. Mauchly and Eckert tried to sell their company to IBM, but IBM declined the offer on the advice of their lawyers, fearing antitrust problems since IBM already controlled much of the mechanical calculator and tabulating machine market. Remington Rand agreed to buy the company as a wholly owned subsidiary, and Mauchly and Eckert became employees in their own company.

The UNIVAC contained 5,000 vacuum tubes that generated so much heat that the engineers and technicians worked in their shorts and undershirts during the hot Philadelphia summer of 1950. The UNIVAC was a true von Neumann machine, with 1,000 memory locations, based on mercury-delay lines, each capable of holding twelve digits or characters. Performing an addition operation took approximately half a millisecond, while multiplication took 2 milliseconds and division took 4 milliseconds. A variation of an electric typewriter served as a control console, and data were fed into the machine through magnetic tapes with nickel-coated bronze tapes, since early efforts with plastic tapes failed. Early versions of the UNIVAC used air cooling, while later versions used a water-cooling system to dissipate the heat generated by the vacuum tubes. Grace Murray Hopper was hired away from the Mark I project at Harvard in 1949 as a senior mathematician to program the new machine, and the first machine was finally delivered to the U.S. Census Bureau in the spring of 1951. Hopper remained with a succession of computer companies until the mid-1960s, when the Navy reactivated her and kept her employed until 1986.

As a publicity effort, Remington Rand proposed to the CBS television network that they use a UNIVAC to predict the winner of the November 1952 presidential election. Mauchly worked for months with a statistician to create a program that used returns from the 1944 and 1948 elections to predict the results in key states from early votes. The UNIVAC program predicted early on the election night such an overwhelming landslide for Dwight D. Eisenhower that the network executives refused to broadcast the results and compelled the programmers to change the program to make the race closer. As the true results came in, showing a crushing defeat for the Democrats, the network admitted that the first results had been correct.

Remington Rand built forty-six UNIVACs for government and business organizations and became the world's leading supplier of data processing computers. Remington Rand merged with Sperry Gyroscope in 1955 to become Sperry Rand, which later merged with Burroughs Corporation in 1986 to form Unisys, briefly the second largest computer corporation in the world. Mauchly left Sperry Rand to form a consulting firm in 1959, while Eckert remained with the company for the remainder of his career.

Mauchly and Eckert were honored as the inventors of the electronic digital computer until 1973, when their patent was overturned in federal court and Atanasoff declared the inventor of the electronic digital computer prior to their efforts. In 1980, Mauchly and Eckert shared the IEEE Computer Society Pioneer Award. Even though their credit for the first computer was misplaced, Mauchly and Eckert did successfully develop the UNIVAC as the first commercial electronic computer in the United States. The original UNIVAC I ran for thirteen years before being donated to the Smithsonian Institution. There is an obvious direct line of technological innovation from the ABC computer to the ENIAC, and then to the UNIVAC. While the Z machines, Harvard Mark 1, bombes, and Colossi remain interesting efforts, they did not substantially affect the larger story of the computer.

3

The Second Generation: From Vacuum Tubes to Transistors

THE COLD WAR

The greatest war in history, World War II, was fought on battlefields and in the laboratory. Never had a war been so dependent on research in science and technology. Among the inventions that poured forth, many of them created through the understanding of science, were radar, computers, jet airplanes, short-range missiles, and the atomic bomb. The Cold War developed in the late 1940s, a global ideological struggle between communism and centrally controlled markets, led by the Soviet Union, and democracy and free markets, led by the United States and its Western European allies. Actual combat encounters between the two superpowers proved to be rare, such as when reconnaissance aircraft were shot down, and quickly concealed in order to not escalate the situation. Other nations instead served as proxies, fighting ideologically based surrogate wars.

Both antagonists realized how important science and technology were to their efforts, and funding levels for research into science and technology continued at unprecedented levels during the Cold War. The governments of the United States, the Soviet Union, and their allies recruited their best and brightest to serve in defense-related research and development. Scientists and engineers developed more advanced computers, computer networks, the

Internet, better medicines, better alloys, industrial ceramics, and technologies with no civilian use, like the neutron bomb.

In the United States, where most computer advances occurred, the military, space exploration, and other efforts by the federal government (especially the nuclear arms program) persistently challenged the emerging computer industry to make computers smaller, more reliable, faster, and more capable. Without this Cold War–induced spending, computer technology would have developed more slowly.

PROJECT WHIRLWIND AND SAGE

During World War II, pilots trained in mechanical trainers before getting in real airplanes. Each airplane model required its own unique trainer. The Navy wanted a single trainer that could be used for many models of airplanes and contracted with the Servomechanisms Laboratory at the Massachusetts Institute of Technology (MIT) to examine the feasibility of such a project. Jay W. Forrester (1918–), a bright young electrical engineering graduate student raised on a cattle ranch in Nebraska, took charge of the project. The Navy envisioned the new flight simulator using servomechanisms to move the simulator and control the flight instruments, all directed by an analog computer that could be reprogrammed to simulate each model of aircraft.

As Forrester worked on the project, he recognized that the simulator was a real-time system, where the computer must respond to real-world input within a set amount of time, and decided that an analog computer would react too slowly to make the simulator work. When Forrester learned about the digital computer projects then going on, especially the ENIAC project, he determined that a electronic digital computer was necessary to make the simulator work. The Navy agreed with Forrester and continued to fund the expanded project, which became named Project Whirlwind in 1946. Forrester and the Navy also realized that the digital computer could be used in many other applications besides flight simulators. The project grew so much in size in the late 1940s that the Navy became concerned about burgeoning costs and wanted to scale back Whirlwind.

The newly formed U.S. Air Force faced its own difficult problem, fearing that Soviet bombers flying in over the Arctic regions would drop atomic bombs on the United States during a war. Radars could detect the bombers, but how could they mesh command and control of all the radars and fighters into a single system to effectively counter the hypothetical Soviet attack? The Air Force turned to MIT, which established the Lincoln

Laboratory to build the Semi-automatic Ground Environment (SAGE) system. Needing a digital computer for SAGE, the Air Force took over funding the Whirlwind project from the Navy, and the Whirlwind engineers designed the FSQ-7 computer. The original purpose for Project Whirlwind, the flight simulator, was never built.

The Air Force contracted with International Business Machines (IBM) to manufacture the AN/FSQ-7 computers for SAGE, giving that company an important opportunity to continue to develop their expertise in electronic digital computers. At one point, one out of every five employees at IBM worked on the SAGE system. The AN/FSQ-7 computers were also the largest computers ever built, each containing over 49,000 vacuum tubes, weighing 250 tons, and occupying a three-story building. The SAGE software eventually totaled over a million lines of programming code and at one point, over half of the programmers in the United States worked on this single project. By 1953, SAGE could simultaneously track forty-eight aircraft, and by 1963 the entire SAGE system of twenty-three direction centers, using twenty-four AN/FSQ-7 computers, was deployed, having cost about $8 billion. The SAGE system continued to function until 1983.

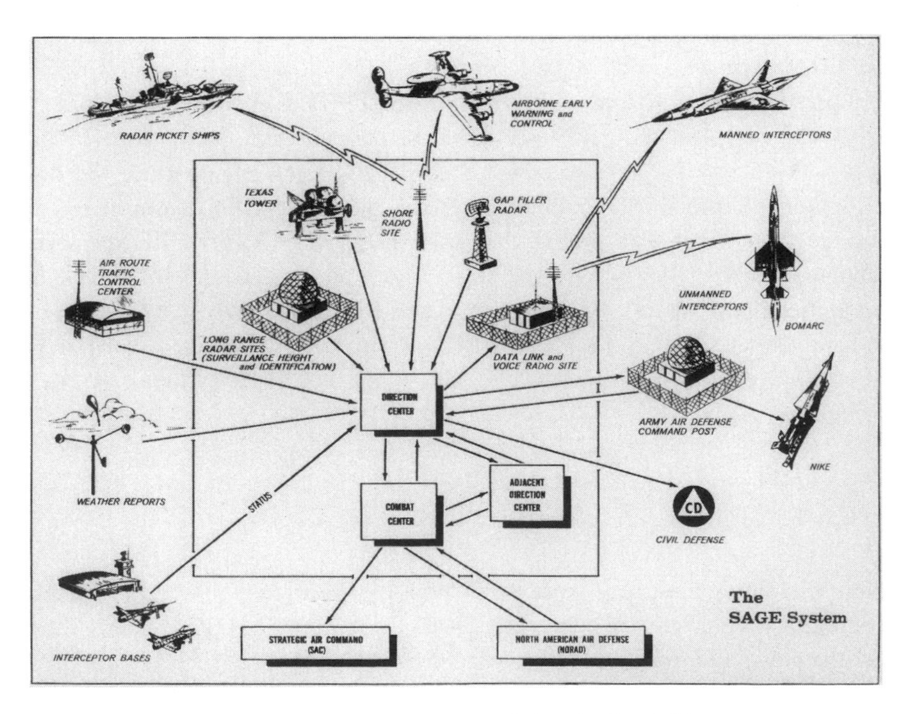

Design of the complex SAGE system. Courtesy of the Charles Babbage Institute, University of Minnesota, Minneapolis.

As with other digital computer projects, a key problem was how to store data in the computer. Forrester and the Whirlwind engineers looked at mercury-delay lines, worked on using electrostatic storage tubes, and eventually perfected magnetic-core memory in 1953. This type of memory stored each bit in a tiny ring of ferrite material, about the size of a pinhead, which retained its on or off binary value magnetically even if electrical power was turned off. This memory technology became the standard memory used in computers until integrated circuits offered a cheaper and faster alternative in the late 1960s.

The SAGE system is considered one of the great successes in early computer history, an example of systems engineering that combined numerous social and technological factors into a functioning whole. SAGE computers used telephone lines to communicate from computer to computer and from computer to radar systems, an early form of modems and networking. SAGE operators used video terminals to track information from the computers and radars, and the light pen was invented as an input device for the operators. The largest real-time computer program written up to that time provided access for simultaneous SAGE operators.

While the American military worked on SAGE, computers were also being developed by the main adversary of the Americans. The Soviet Union built their first electronic digital computer in 1950 at the Kiev Institute of Electric Engineering under the direction of S. A. Lebedev (1902–1974), which ran fifty instructions a second and used a memory of thirty-one 16-bit words. After moving to Moscow, Lebedev proceeded to supervise the development of the BESM (which stood for "large electronic computer" in Russian) series of computers that used magnetic drums and magnetic tapes for storage. By and large, Soviet computers were copies of American computers, purchased illegally through third-party countries, since during most of the Cold War, the United States forbade the export of computers to communist companies, considering computers to be as lethal as arms or munitions.

TRANSISTORS

Early computer designers were tormented by the nature of vacuum tubes, constantly struggling to deal with excess heat, their bulky size, and their penchant for failure. The room-sized dimensions of computers came directly from efforts to cope with vacuum tubes. Salvation came from Bell Telephone Laboratories in Murray Hill, New Jersey, where the theoretical physicist John Bardeen (1908–1991) and experimental physicist Walter H.

Brattain (1902–1987) invented the point-contact transistor, made out of germanium, a semiconductor material, in 1947. The physicist William B. Shockley (1910–1989) was the team leader for the project, but was not involved in the initial invention. Shockley took the invention and made the junction transistor, which became the basis for later commercial development. Transistors acted as a switch to turn on or off the flow of electricity and as an amplifier to the current. The American military immediately recognized the long-term value of transistors and provided additional funding to further develop the invention.

Though the three Bell scientists shared the 1956 Nobel Prize in Physics for their invention, rivalries among the three drove them apart. Shockley returned to Palo Alto, California, where he had grown up, in 1956 and founded Shockley Transistor. Under the influence of nearby Stanford University, the surrounding towns already made up a nascent center of technological innovation. Shockley's decision to locate there eventually led to the area becoming known as Silicon Valley. Shockley intended to create commercial transistor products, but his abusive management style alienated his employees. A year later eight of his employees fled to form their

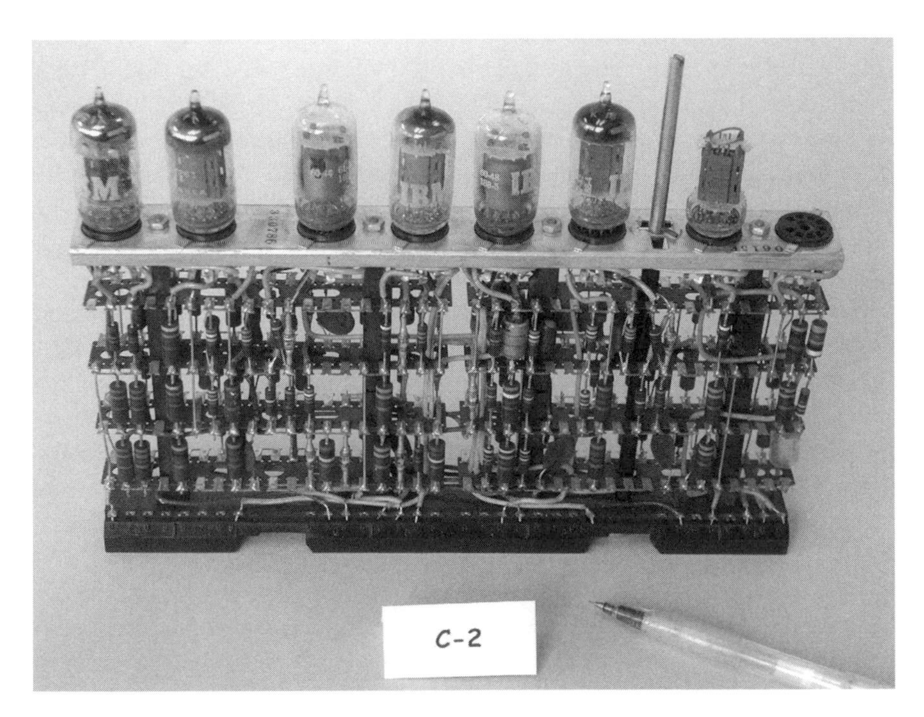

Photograph of an IBM electronic component combining both transistor and vacuum tube technology. Courtesy of Weber State University.

own company, Fairchild Semiconductor. After Shockley's company faded away, he turned to the study of genetics and became infamous for promoting racist views of human inheritance.

By the late 1950s, the transistor became a useful commercial product, creating the second generation of computer hardware, rapidly replacing vacuum tubes in computers and other electronic devices because transistors were much smaller, generated less heat, and were more reliable. Defense-related projects and space-related projects in the United States, as part of the Cold War, became a major driver for computer-related innovation in both hardware and software.

COMMERCIAL COMPUTING AND IBM

Early computers exclusively focused on mathematical processing for scientific and engineering applications. A large data processing business already existed, having transformed how businesses worked in the first half of the twentieth century. Insurance companies, banks, governments, and other organizations who processed large amounts of data relied in mechanical calculators, punched cards, and tabulators. The UNIVAC electronic computers promised to revolutionize this market. IBM dominated the market for punched-card equipment, and in 1951 the company changed its sales predictions to include a substantial emerging commercial market for electronic computers. IBM already had substantial electronic computing experience through the work on the Harvard Mark I. The contract to produce the AN/FSQ-7 computers for the SAGE system also helped IBM catch up on the technology and even develop leading-edge technologies.

Approximately ten large computer companies emerged in the 1950s, all competing to dominate the new market. The United States, flush with postwar prosperity, provided most of the customers for new electronic computers, since other industrialized nations were still rebuilding their basic infrastructure after the devastation of World War II. The new computer companies were usually preexisting electronics manufacturers or business machine manufacturers expanding into a new area, and included General Electric, RCA, Raytheon, Honeywell, Burroughs, Remington Rand, Monroe, and Philco. IBM had an advantage over other companies because of their highly motivated sales force that was accustomed to developing complete data processing solutions for their customers, not just selling or leasing equipment.

In the early 1950s, as commercial companies began to build production computers, following the example of the UNIVAC, there were many one-of-a-kind computers built, such as the IAS by John von Neumann at the

Institute for Advanced Study located at Princeton University. The National Applied Mathematics Laboratories, part of the National Bureau of Standards, built the Standards Eastern Automatic Computer (SEAC) in their Washington, D.C., office, and the Standards Western Automatic Computer (SWAC) in their Los Angeles office, both in 1950. The National Physical Laboratory in Teddington, Middlesex, England, built their Pilot Automatic Computing Engine (ACE) in 1950, designed by Alan Turing. IBM delivered the Naval Ordnance Research Calculator (NORC) for the U.S. Navy's Bureau of Ordnance in 1954. Though the NORC was built for its actual cost, with the company not charging extra for profits, IBM benefitted from technical advances that they applied to later computers. All these computers were based on vacuum tube technology and were designed for scientific calculations, not data processing.

While working on both the NORC and AN/FSQ-7 computers, IBM also canvassed the market for further government, defense-related, and aviation-related customers. They found thirty customers willing to sign letters of intent, and began work on what became known as the IBM 701. The 701 drew heavily on the design of the Princeton IAS, and used Williams' tube electrostatic memory units developed at Manchester University to achieve a total capacity of 4,096 36-bit words. The original memory units were visible through glass doors, allowing computer operators to see the contents of each memory location as dots on the cathode ray tubes. A photographer at the official unveiling of the computer in 1952 used a flashbulb, which promptly reset the tube-based memory with random bits. The IBM engineers built later memory units with darkened glass covers.

IBM brought their strategy of leasing equipment, rather than selling it, from their punched-card business into their electronic computer business. The 701 leased for $8,100 a month. Because the 701 proved more expensive than initially projected, only nineteen customers leased the machines. Douglas Aircraft Company purchased two to help design such noted aircraft as their DC-6B, DC-7, and DC-8 aircraft, as well as the A-3D Skywarrior and A-4D Skyhawk aircraft for the U.S. Navy. Lockheed Aircraft Company also purchased two IBM 701 computers. The users of the IBM 701 formed a group called SHARE, which became an informal mechanism for exchanging programs among themselves and pressuring IBM to create more software products. Already everyone recognized that one of the advantages of common machines was that programs did not have to rewritten for each individual computer. User groups for computers from other manufacturers also became common. These user groups provided a way for customers to unite and encourage manufacturers to implement preferred features to develop particular types of software. Chapter meetings and annual conferences became a way for the manufacturer to communicate with customers, and for people with job skills on a particular computer model to find a job with another company with that same computer model.

IBM followed up the 701 with their 704 computer. Brought to market in 1954, the 704 was the first mass-produced computer to use magnetic core memory and offer built-in floating-point hardware for handling decimal numbers. IBM also improved the 704 over the 701 by providing a set of seventy-five basic instructions for the 704, whereas the 701 only had thirty-three instructions. The greater number of instructions made it easier for programmers to do more complex operations. The 704 also used a rotating magnetic drum for secondary memory.

The IBM 702, which completed development a year after the 704, was not designed as a fast scientific calculator, as earlier computers were, but for business data processing. To facilitate data processing, which emphasized transactions of characters and numbers, the memory of the 702 was oriented toward storing characters. The 702 also included a new tape drive that successfully used plastic-based magnetic tapes and prompted the entire industry to switch away from the metal-backed tapes that the UNIVAC had pioneered. After only fourteen machines, the 702 was superceded by the 705, which IBM continued to upgrade until it reached 80,000 characters of memory in 1959, and the company leased 180 of the machines.

IBM also introduced the 650 Magnetic Drum Data Processing Machine in 1954, designed as a business data processing machine midway between the punched-card systems still in wide use and the processing power of the IBM 702. The 650 used a rotating drum to store 2,000 ten-digit words, which contained both instructions and data. Though the drum spun at 12,500 revolutions per second, the drum took about 5 milliseconds to complete a rotation. Many computer instructions took only about 3 milliseconds to execute, so when instructions were laid sequentially across the drum, about 2 milliseconds were wasted during every execution cycle. The solution was to carefully stagger the instructions of a program across the drum to minimize the latency between instructions. The 650 leased for $3,250 a month, and IBM eventually sold 2,000 of these machines, earning more money than from their entire 700 series. With a canny understanding of market dynamics, IBM offered deep discounts on 650 computers to institutions of higher education, as long as the college or university began to teach programming. Of course, the students learned to program IBM computers. The IBM 650 established IBM's lead in the electronic computer industry; and in 1955, even orders for just their more flashy 700 series computers exceeded orders for all Sperry Rand UNIVACs.

The economics of the time meant that computers were expensive, and anything that made their use more efficient was desirable. A major problem for efficiency came from the fact that peripheral devices were so much slower that the main central processing unit (CPU), and the CPU was often idle while a tape was being read or written, or while data was being sent to a line printer. IBM introduced their smaller 1401 computer, whose sole task was writing data to a magnetic tape or copying data from a tape and sending it to a line printer, and eventually sold over 12,000 IBM 1401 model computers. IBM complemented the 1401 by developing their 1403 printer that produced 600 lines per minute. The printer used a horizontal chain with characters on it that moved rapidly back and forth while hammers slammed the characters on the chain into an ink ribbon, leaving a printed character on the paper. IBM eventually sold more than 23,000 of these printers, keeping them available for sale up until 1983.

General Electric (GE) got its start in the computer industry by building specialized computers for the military. In the mid-1950s, an executive at GE became aware of an commercial opportunity at Bank of America (BoA). Concerned about their ability to efficiently handle the millions of checks that flowed through their banks every year, BoA contracted with the Stanford Research Institute (SRI) to build a prototype machine that processed checks and recorded the necessary accounting information. SRI developed magnetic ink character recognition (MICR) so that magnetic characters imprinted on the checks could be read by the check readers. Burroughs also developed a technology using fluorescent dots, and IBM preferred to use bar codes, to solve the problem of how to get the check reader to quickly read the checks. The SRI machine, finished in 1955, included 8,000 vacuum tubes, 34,000 diodes, and a million feet of copper wire. The machine was hardwired, not using any form of software or programming.

General Electric won the bid on the contract to build thirty-six production models of this machine at a cost of $31 million. The natural choice to win the bid, IBM, did not bid, but tried instead to buy the idea from BoA and SRI so that they could develop it into a general solution to sell to many banks. BoA chose not to sell, and GE went to work. The proposed Electronic Recording Machine Accounting (ERMA) computer was completely redesigned, using 5,000 transistors, 15,000 diodes, and 4,000 resisters, turning it from a hardwired computer into a stored-program computer. An early programming language, Intercom 100, was developed just for those computers. In late 1958, the first computer was placed in a bank, able to only process 100 banking transactions per day. After further refinement, in just three months, the computer achieved the 55,000 transactions per day required by the contract. The check sorter and check reader, developed by NCR, could even handle checks that had been crumpled and stepped on before being smoothed out by hand and set in the sorter. GE renamed ERMA the GE 100 and entered the commercial computer industry in earnest, though the company leadership always remained leery of competing with IBM head to head.

As always, American defense spending remained an important driver in the first three decades of electronic computers. In 1955, the Los Alamos Scientific Laboratory of the U.S. Atomic Energy Commission requested a computer 100 times faster than any then in existence, and IBM decided to take up the challenge. Instead of building a one-of-a-kind machine like the previous NORC, IBM designed the computer to meet the Los Alamos requirement and to become a new computer, the IBM 7030. The project was appropriately named Stretch. The new computer used transistors, faster magnetic core memory, and the pipelining of instructions. Pipelining is when electronic circuitry in the CPU is designed to not only execute the current instruction, but also begin decoding and processing subsequent instructions at the same time. The 7030 decoded and partially processed the five instructions beyond the current instruction, ready to throw away the extra work if an earlier instruction proved to be a branch or to make some change that made the work of the later instructions invalid. The 7030 also had a memory controller that prefetched data from memory locations before the CPU asked for them, anticipating the needs of the program. IBM pioneered random access disk drives, and the drive for the 7030 was the first drive to include more than a single read/write arm in the same movable system, leading to much greater storage capacity in hard drives.

The 7030 was the most complex computer ever built up to that time, and IBM engineers used programs on an IBM 704 to simulate the features of the system, especially the pipelining features. Computers were now being used to design new computers. Though the Stretch project did not quite meet its goal of a hundredfold increase in performance, the first 7030 was delivered to Los Alamos in 1961 and seven others were produced for customers in England, France, and the United States. One of the machines became the core of the Harvest machine for the U.S. National Security Agency, the government agency tasked with code breaking.

The pace of technological innovation forced IBM to come up with ever newer machines. The 709 was first produced in 1958 as a successor to the 701 and 704 series. IBM took care to make the 709 downwardly compatible, which meant that programs written for the early machines could run without modification on the newer machine. Philco Corporation introduced the Transac S-2000 in 1958, the first commercial computer built with transistors instead of vacuum tubes. A new model of the UNIVAC was also being prepared using transistors. IBM quickly reacted by building a new version of the 709, the 709TX, using transistor technology. The 709TX was designed for use in the Ballistic Missile Early Warning System (BMEWS), a complement to the SAGE system. Whereas SAGE detected the previous threat from aircraft, the BMEWS sought to detect the new threat of nuclear-tipped intercontinental ballistic missiles (ICBMs). IBM converted the 709TX into their 7090 computer for commercial customers, which ran five times faster than the 709 and had much greater reliability because of the use of transistors instead of vacuum tubes.

By the end of the 1950s, IBM had a wide range of computers to meet the different needs of the scientific and data processing industries. This large number of systems confused the marketplace and fragmented the technical efforts at IBM, yet IBM had come to dominate the industry to such an extent that people took to calling the situation Snow White and the Seven Dwarfs, where IBM was Snow White and a slew of smaller companies, not always seven in number, tried to survive in IBM's shadow. IBM's industrial designers also created a light blue design for the computer cabinets, earning the company the nickname "Big Blue." The company employees also took to wearing blue suits and white shirts. Despite its dominance of the electronic computer industry, in 1959 the majority of IBM's revenue still came from its punched-card business, not computers. Overseas sales for IBM were also important, with 20 percent of company revenue in 1960 coming from outside the United States. This increased to 35 percent in 1969, 54 percent in 1979, and 61 percent in 1990. The American computer industry provided computer hardware and software for the rest of the world, usually pushing local companies to the margins.

Only Sperry Rand and their UNIVAC computers were successful enough to continue to directly compete with IBM. Whereas IBM had built the 7030 in their Stretch project, Sperry Rand build the Livermore Atomic Research Computer (LARC) for the Lawrence Radiation Laboratory of the Atomic Energy Commission in California. The LARC was comparable to the 7030, though it used high-speed drums rather than the newer disk-drive technology. After delivering the LARC in 1960, Sperry Rand built a second one for the U.S. Navy Research and Development Center in Washington, D.C.

The Stretch project and the LARC were supercomputers compared to contemporary machines. The third supercomputer project came from Ferranti in England, where Tom Kilburn directed a team to built the Ferranti Atlas. The Atlas pioneered two important technologies: virtual memory and some aspects of time sharing. The Atlas was designed to use a memory space of up to a million words, with each word 48 bits long. No one could afford to put that much magnetic core memory in a machine, so the Atlas had actual core memory of only 16,000 words. A drum provided 96,000 more words. The operating system of the Atlas swapped memory from its magnetic core memory to the drum and back as needed in the form of pages, providing the illusion of more memory via this virtual memory scheme. The Atlas also was designed to be a time-sharing computer so that more than one program at a time could be run. To implement this time sharing, the idea of extracode was developed, which is similar to what are now called system interrupts. These two ideas were adopted in all later operating systems of any sophistication. As with the other supercomputers, the Atlas was not a commercial success, since only three were built.

PROGRAMMING LANGUAGES

The first programs were created by rearranging the plugs and wiring on the computer. The von Neumann idea of using memory to hold the program instructions and data, not just the data, required that another method of entering the program be invented. Punched cards, paper tape, and magnetic tape were soon adopted. The first programs were written in straight binary, also called machine code, and were difficult to write and debug. What programmers of the time called "automatic programming" soon emerged, mainly in an effort to reuse binary code that performed common functions, such as floating-point arithmetic, which is calculations using decimal numbers. As these automatic programming systems grew more complex, they frustrated programmers because the resulting code was not as efficient and ran five to ten times slower than programs written from scratch.

The economics of computing costs had already emerged in the early 1950s, where programmer and computer operator salaries became the major cost of a computer center. A common feature of high technology is that people cost more than the machines that they use. Managers found that when computers were used for both development and production activities, most computer time was taken up by programming and debugging activities rather than getting production work done. Anything that increased the efficiency of programmers helped the productivity of the computer center, but what if the programs created via automatic programming ran so slow that the economic advantage in their method of creation was lost during the years that the programs ran?

Programmers attacked the problem by creating higher level computer languages. Some early examples of these languages were algebraic compilers, and languages with such names as Short Code, Mathe-Matic, Speedcoding, and Autocode. The IBM 704 had floating-point logic built into its circuits so that programmers no longer had to write code to manipulate decimal numbers, and an IBM employee, John Backus (1924–), saw an opportunity. He proposed in 1954 to create a new programming language that made it easier to write computer programs but was efficient enough to compete with hand-coded binary programs. IBM gave Backus a small team of programmers, and they set out to create FORTRAN (FORmula TRANslator). The team completely focused on creating efficient object code and literally made up the language as they went along, in contrast to most later programming languages, which are planned out before development. The team also decided to ignore any blanks in the source code, since blanks always caused a lot of problems for people running keypunches.

The FORTRAN project took longer than anticipated and was not completed until April 1957. The initial release of the compiler came in a deck of 2,000 punchcards. The language became popular with users of the IBM 704 and the language rapidly went through new versions, adding new features, especially features that helped with debugging. FORTRAN II came out in the spring of 1958, followed by FORTRAN III and FORTRAN IV, with the FORTRAN 66 definition coming out in 1966. The language was also ported to other computer systems besides the IBM family of computers and served as the basic programming language for scientific and engineering applications for decades. FORTRAN was simple enough that scientists and engineers could learn to write programs themselves.

One of the more unfortunate stories of FORTRAN programming occurred in the early space program. The U.S. government designed the Mariner spacecraft series to be the first interplanetary spacecraft, carrying cameras and scientific instruments, powered by solar panels, and radioing their findings back to Earth. An Atlas Agena B rocket launched the first Mariner spacecraft on July 22, 1962. A guidance signal from the ground directed the rocket's ascent. When that signal failed, an onboard computer took over. A programming mistake in the FORTRAN program running on the onboard computer left out a comma from the computer program, causing the rocket to veer to the left. A range safety officer detonated an explosive charge within the rocket to prevent any harm to people watching the launch. Later Mariner spacecraft, with the bug fixed, successfully visited the planets Venus, Mars, and Mercury.

While FORTRAN served the scientific and engineering segment of computer users, business users were becoming ever more important. Business users often wrote programs that processed data in the form of transactions, for instance, reservations for airlines, or checks being processed and cleared.

The U.S. Department of Defense also had a need for data processing programs and sponsored the creation of COBOL (COmmon Business-Oriented Language). Grace Hopper (1906–1992) helped lead the effort to create this language and remained heavily involved with Navy computer programming efforts. The Navy waived their mandatory retirement age and she eventually retired in 1986, after attaining the rank of rear admiral. Some programmers considered COBOL too wordy, but other programmers lauded the readability of using complete English words like MULTIPLY and READ, rather than the terse syntax of FORTRAN.

FORTRAN and COBOL are considered third-generation computer languages, following the assembly languages that make up the second generation, and the binary machine code of the first generation. The historical generations of programming languages do not correspond with the historical generations of computing hardware. For instance, COBOL and FORTRAN were written on second-generation hardware based on transistors. IBM also developed Report Program Generator (RPG), a simple language used to quickly generate simple business and accounting applications. The language was designed to mimic punched-card machines, so that people already trained in using plugboards on punched-card machines could easily transfer their arcane expertise to the computer. Other languages also emerged at the time, such as List Processing (LISP) in 1958 and Algorithmic Language (ALGOL) in 1960.

SOFTWARE SYSTEMS

As the industry developed during the 1950s, IBM began to develop suites of software aimed at particular industries, such as banks, manufacturing, or insurance. IBM gave away the software to those industries as an incentive to buy computers, computer peripherals, and services. Other computer companies did not have the deep financial resources of IBM and concentrated on one or two classes of customers, developing software to give away so that they could compete with IBM in that more restricted arena. Burroughs and GE concentrated on banking, NCR concentrated on retailing, and Control Data Corporation concentrated on scientific computing.

In 1955, two programmers at IBM left to form the first computer software services company, the Computer Usage Company (CUC). The company's first project simulated the radial flow of fluids in an oil well for the California Research Corporation. Other contract programming projects followed. The RAND Corporation, a think tank owned by the U.S. government, created a subsidiary, Systems Development Corporation (SDS), in

1957 to write computer programs for the SAGE air defense project. Computer Sciences Corporation (CSC) was formed in 1959, and their first contract was to develop a business-language compiler for Honeywell. CSC initially worked on writing systems software for computer manufacturers, but later focused on providing computer contracting services to the federal government and military. Software contracting firms also emerged in Europe, usually a few years later than American examples. For example, in 1958, Banque de Paris and Marcel Loichot formed the Sema company in France.

Until the 1960s, all software—with the exception of some system software, such as operating systems, interpreters, compilers, and system utilities—was developed for individual applications. All of the early computer software companies relied on contract programming for their income, since a software market had not yet developed to sell their software independently. Many of these early software systems automated previous manual business processes or business processes that had been partially automated by punchcard systems.

An example of an early manual data processing system, the Reservisor system at American Airlines, was used to arrange airline reservations. As the jet age began, causing explosive growth in the airline industry, American Airlines planned to expand the Reservisor system, but realized that their manual methods would no longer function on the scale required. IBM became involved in finding a solution for American Airlines. Realizing how long a computerized system would take to develop, IBM created a temporary system based on punched-card processing for use in the late 1950s. The computerized SABRE system, began in 1957, was finally completed in 1964, based on two IBM 7090 mainframes running a million lines of code. Including the largest online storage system up to that time, 800 megabytes in size, SABRE supported 1,100 travel agents around the country connected to the system via terminals. SABRE has continued run to this day, standing as an example of a large real-time transaction-processing system with sophisticated measures to avoid any downtime. SABRE became the basis for the well-known web site, http://www.travelocity.com, created in the 1996 with SABRE providing the background data processing engine and the new web technology, providing a way to directly reach customers with an easy-to-use interface.

EARLY EFFORTS AT ARTIFICIAL INTELLIGENCE

Early computer pioneers, like the mathematicians Alan Turing (1912–1954) and John von Neumann (1903–1957), intentionally developed electronic computers as the first step toward the creation of genuine thinking machines. In 1950, Turing created the "Turing Test," which proposed how to test an intelligent machine. The test required a human to have a conversation with a computer via a teletype terminal, and if the human could not determine if the answers came from a human or a computer, then artificial intelligence had been achieved. Of course, this test would have to be conducted many times with many people before that final conclusion was reached. This is now called the "Turing Test," and no computer or program has come close to passing it. While contemporary researchers no longer commonly accept the "Turing Test" as an absolute criterion, it remains a good indicator of the goals of these early pioneers.

The mathematician Marvin Minsky (1927–) became one of the pioneers of research in artificial intelligence. Minsky was born in New York, where his father was an eye surgeon and his mother was a Jewish activist. Minsky attended private schools where his gifts were recognized, and after brief service in the U.S. Navy, Minsky entered Harvard University in 1946. Initially majoring in physics, he expanded his interests into psychology after becoming fascinated with how the mind worked, and finally graduated in 1950 with B.A. in mathematics. Minsky moved to Princeton University, where a colleague and he built in 1951 the Stochastic Neural-Analog Reinforcement Computer, or Snarc. Made out of 400 vacuum tubes, the machine was an early attempt at creating a learning system based on neural nets like those the human brain used. Minsky earned a Ph.D in mathematics from Princeton in 1954, then returned to Harvard as a junior fellow, where he worked on microscopes and patented a scanning microscope.

Minsky and John McCarthy (1927–) organized a two-month summer workshop at Dartmouth College in 1956, where the term "artificial intelligence" (AI) was first coined. Artificial intelligence research became defined as the effort to create computer hardware and computer software that behave as humans do and that actually think. In 1958, Minsky and McCarthy joined the faculty of the Massachusetts Institute of Technology (MIT). A year later, Minksy and McCarthy founded the MIT Artificial Intelligence Project, which became the Artificial Intelligence Laboratory in 1964. Learning in 1956 that IBM planned to donate one of their 704 computers to MIT and Dartmouth, McCarthy conceptually developed a new programming language for use in AI research before the computer even arrived. His algebraic list-processing language became LISP in 1958 and is still used in AI research.

The first two decades of AI research were dominated by researchers at the Massachusetts Institute of Technology (MIT), Carnegie Tech (later renamed Carnegie Mellon University), Stanford University, and International Business Machines (IBM). European efforts and Japanese efforts later became important. The history of AI has been characterized by a series of theories that showed initial promise when applied to limited cases, leading to optimistic declarations that intelligent machines were just around the corner, and disappointment as the theories failed when applied to more difficult problems.

4

The Third Generation: From Integrated Circuits to Microprocessors

INTEGRATED CIRCUITS

Jack S. Kilby (1923–) grew up in Great Bend, Kansas, where he learned about electricity and ham radios from his father, who ran a small electric company. Kilby desperately wanted to go to the Massachusetts Institute of Technology, but failed to qualify when he scored 497 instead of the required 500 on an entrance exam. He turned to the University of Illinois, where he worked on a bachelor's degree in electrical engineering. Two years of repairing radios in the Army during World War II interrupted his education before he graduated in 1947. He moved to Milwaukee, Wisconsin, where he went to work for Centralab. A master's degree in electrical engineering followed in 1950. Centralab adopted transistors early, and Kilby became an expert on the technology, though he felt that germanium was the wrong choice for materials. He preferred silicon, which could withstand higher temperatures, though it was more difficult to work with than germanium. After Centralab refused to move to silicon, Kilby moved to Dallas to work for Texas Instruments (TI) in 1958.

Texas Instruments had been founded in the 1930s as an oil exploration company and later turned to electronics. TI produced the first commercial silicon transistor in 1954 and the first commercial transistor radio that same year. In his first summer at the company, Kilby had no vacation days accrued when everyone else went on vacation, so he had a couple of weeks of solitude at work. TI wanted him to work on a U.S. Army project to build Micro-Modules, an effort to make larger standardized electronics modules. Kilby thought the idea ill-conceived and realized that he had only a short time to come up with a better idea.

Transistors had transformed the construction of computers, but as ever more transistors and other electronic components were crammed into smaller spaces, the problem of connecting them together with wires required a magnifying glass and steady hands. The limits of making electronics by hand became apparent. Making the electronic circuitry larger by spacing the components farther apart only slowed down the machine because electrons took more time to flow through the longer wires.

Transistors were built of semiconductor material and Kilby realized that other electronic components used to create a complete electric circuit, such as resistors, capacitors, and diodes, could also be built of semiconductors. What if he could put all the components on the same piece of semiconductor material? On July 24, 1958, Kilby sketched out his ideas in his engineering notebook. The idea of putting everything on a single chip of semiconductor later became known as the *monolithic* idea or the integrated circuit.

By September, Kilby had built a working integrated circuit. In order to speed up the development process, Kilby worked with germanium, though he later switched to silicon. His first effort looked crude, with protruding gold wiring that Kilby used to connect each component by hand, but the company recognized an important invention. Kilby and TI filed for a patent in February 1959.

In California, at Fairchild Semiconductor, Robert Noyce (1927–1990) had independently come up with the same monolithic idea. Noyce graduated with a doctorate in physics from the Massachusetts Institute of Technology in 1953, then turned to pursuing his intense interest in transistors. Noyce worked at Shockley Transistor for only a year, enduring the paranoid atmosphere as the company's founder, William B. Shockley (who had been part of the team that invented the transistor in 1947), searched among his employees for illusionary conspiracies. Noyce and seven other engineers and scientists at the company talked to the venture capitalist who provided the funding for the company, but could obtain no action against Shockley, a recent Nobel laureate. Shockley called the men the "traitorous eight" when Noyce and his fellow rebels found financing from Fairchild Camera and Instrument to create Fairchild Semiconductor.

Fairchild Semiconductor began to manufacture transistors and created a chemical planar process that involved applying a layer of silicon oxide on top of electronic components to protect them from dust and other contaminants. This invention led Noyce to develop his own idea in January 1959 of the integrated circuit, using the planar process to create tiny lines of metal between electronic components to act as connections in a semiconductor substrate. After Texas Instruments and Kilby filed for their patent, Noyce and Fairchild Semiconductor filed for their own patent in July 1959, five months later. The latter patent application included applying the chemical planar process in it.

Kilby and Noyce always remained friendly about their joint invention, while their respective companies engaged in a court fight over patent rights. Eventually the two companies and two engineers agreed to share the rights and royalties, although the U.S. Court of Customs and Patent Appeals ruled in favor of Fairchild in 1969. The two companies agreed, as did the two men, that they were coinventors. Kilby later received half of the 2000 Nobel Prize in Physics for his invention, a recognition of the importance of his invention. Noyce had already died, and Nobel prizes are not awarded posthumously. The other half of the Nobel Prize was shared by the Russian physicist Zhores I. Alferov (1930–) and German-born American physicist Herbert Kroemer (1928–) for their own contributions to semiconductor theory.

The commercial electronics industry did not initially appreciate the value of integrated circuits (also called microchips), believing them too unreliable and too difficult to manufacture. However, both the National Aeronautics and Space Administration (NASA) and the American defense industry recognized the value of microchips and became significant early adopters of microchips, proving that the technology was ready for commercial use. The U.S. Air Force used 2,500 TI microchips in 1962 in the onboard guidance computer for each nuclear-tipped Minuteman intercontinental ballistic missile. NASA used microchips from Fairchild in their Project Gemini in the early 1960s, a successful effort to launch two-man capsules into orbit and to rendevous between a manned capsule and an empty booster in orbit. NASA's $25 billion Apollo Project to land a man on the moon became the first big customer of integrated circuits and proved a key driver in accelerating the growth of the semiconductor industry. At one point in the early 1960s, the Apollo Project consumed over half of all integrated circuits being manufactured.

NASA requirements and military requirements also led to advances in the creation of fault-tolerant electronics. If a business computer failed, a technician could repair the problem and the nightly accounting program could continue to run; if the computer on the Saturn V rocket failed, then the astronauts aboard might die. Electronics were also hardened to survive the shaking from a rocket launch and exposure to the harsh vacuum and temperatures of space.

Similar efforts also led to the development of the idea of software engineering, which is applying sound engineering principles of verification to the development of programs to ensure that they are reliable in all circumstances.

Commercial industries began to appreciate the value of integrated circuits when Kilby and two colleagues created the first electronic calculator using microchips in 1967. The calculator printed out its result on a small thermal printer that Kilby also invented. This was the first electronic calculator small enough to be held in a hand and sparked what became a billion-dollar market for cheap, handhand calculators, quickly banishing slide rules to museums. Integrated circuits became the main technology of the computer industry after a decade of development, creating a third generation of computer technology (following the generations based on vacuum tubes and transistors).

In 1964, Gordon E. Moore (1929–) noticed that the density of transistors and other components on integrated chips doubled every year. He charted this trend and predicted that it would continue until about 1980, when the density of integrated circuits would decline to doubling every two years. Variations of this idea became known as Moore's Law. Since the early 1970s, chip density on integrated circuits, both microprocessors and memory chips, has doubled about every eighteen months. From about fifty electronic components per chip in 1965, 42 million electronic components were placed on an individual chip in 2000. An individual transistor on a chip is now about 90 nanometers (90 billionths of a meter) in size. At times, different commentators have predicted that this trend would hit an obstacle that engineers could not overcome and slow down, but that has not yet happened, though the electronic components on integrated circuits are being packed so close that within a decade engineers fear that quantum effects will begin to substantially affect that ability of Moore's Law to remain true.

Moore also pointed out another way to understand the growth of manufactured semiconductor material. From the beginning, every acre of silicon wafer has sold for about a billion dollars; the number of transistors and other electronic components on the ship have just become more dense to keep the value of that acre roughly constant. The following list shows the growth of microchips by the date first created, the technology used, and how dense each integrated circuit could be.

1960	Small-scale integration (SSI)	Less than 100 transistors
1966	Medium-scale integration (MSI)	100–1,000 transistors
1969	Large-scale integration (LSI)	1,000–10,000 transistors
1975	Very large-scale integration (VLSI)	10,000–100,000 transistors
1980s	Ultra large-scale integration (ULSI)	More than 100,000 transistors
1990s	Still called ULSI	More than 1 million transistors

While the manufacture of integrated circuits is considered to be part of the electronics industry, the industrial techniques used are more like those of the chemical industry. A mask is used to etch patterns into wafers of silicon, creating a large number of integrated circuits in each batch. The key to economic success is getting high yields of mistake-free batches.

MINICOMPUTERS

Ken Olsen (1926–) founded Digital Equipment Corporation (DEC) in 1957, and began manufacturing electronics using the new transistor technology. In 1965, DEC introduced the Programmed Data Processor-8 (PDP-8), the first mass-produced computer based on integrated circuits. The entire PDP-8 fit in a normal packing crate and cost about $18,000. By 1969, the PDP-8 was recognized as the first minicomputer. Minicomputers were not as powerful as what had became known as mainframes, and were usually bought to be dedicated to a small number of tasks, rather than as a general-purpose business data processing computer. Other companies also entered the minicomputer market, including Data General, Prime Computer, Hewlett-Packard, Harris, and Honeywell. By 1970, based on its minicomputers, DEC was the third largest computer manufacturer in the world, behind IBM and Sperry Rand. DEC eventually became the second largest computer company, based on the strength of its PDP series and, later, VAX series of minicomputers.

After the personal computer emerged in the 1970s, minicomputers occupied the middle ground between microcomputers and mainframes. Minicomputers eventually ran sophisticated operating systems, and were heavily used in the engineering, scientific, academic, and research fields. In the 1980s, minicomputers also found their way to the desktop as workstations, powerful single-user machines often used for graphics-intensive design and engineering applications. In the 1990s, minicomputers and workstations disappeared as market segments when personal computers became powerful enough to completely supplant them.

TIMESHARING

Early computers were all batch systems, which is where a program is loaded into a computer and run to completion, before another program is loaded and run. This serial process allowed each program to have exclusive access to the computer, but frustrated programmers and users for two reasons. First, the central processing unit (CPU) lay idle while programs were loaded, which wasted expensive computer time; second, batch processing made it difficult to do interactive computing. In a 1959 memorandum, John McCarthy (1927–), already a founding pioneer in artificial intelligence, proposed to MIT that a "time-sharing operator program" be developed for their new IBM 709 computer that IBM planned to donate to the prestigious school. Christopher Strachey (1916–1975) in the United Kingdom simultaneously and independently came up with the same idea.

By late 1961, a prototype of the Compatible Time-Sharing System (CTSS) was running at MIT. Further iterations of CTSS were created, and in the mid-1960s a version of CTSS implemented the first hierarchical file system, familiar to users today as the idea of putting files into directories or folders to better organize the file system. J.C.R. Licklider (1915–1990) of the Advanced Research Projects Agency, a branch of the Pentagon, was a keen advocate of interactive computing and funded continued work on timesharing. Other American and British projects also researched the goal of getting multiple programs to run in the same computer, a process called multiprogramming. Though only one program at a time can actually run on the CPU, other programs could be quickly switched in to run as long as the other programs were also resident in memory. This led to the problem of how to keep multiple programs in memory and not accidently have one program overwrite or use the memory already occupied by another program. The solution to this was a series of hardware and software innovations to create virtual walls of exclusion between the programs.

Operating system software became much more sophisticated to support multiprogramming and the principle of exclusion. An ARPA-MIT project called Multiplexed Information and Computing Service (or Multics), began in 1965, did not realize its ambitious goals in that it was only a modest commercial success, but became a proving ground for many important multiprogramming innovations. Two programmers who worked on Multics, Dennis M. Ritchie (1941–) and Ken Thompson (1943–) at AT&T's Bell Laboratories, turned their experience into the UNICS operating system. The name stood for UNiplexed Information and Computing Service, a pun on Multics, but later was shortened to UNIX. Originally written in assembly language on a DEC PDP-7 minicomputer, Ritchie and Thompson wanted to port UNIX to a new minicomputer, the DEC PDP-11, and decided to rewrite the operating system in a higher-level language. Ritchie had created the programming language C (a successor to a language called B), and the rewritten UNIX was the first operating system written in a third-generation language. As a government-sanctioned monopoly, AT&T was not allowed to sell any of its inventions outside of the telephone business, so AT&T offered UNIX to anyone who wanted to buy it for the cost of the distribution tapes and manuals, though AT&T retained the copyright. Because it was a full-featured operating system with all the source code included, UNIX became popular at universities in the 1970s and 1980s.

IBM SYSTEM/360

In the early 1960s, IBM had seven mutually incompatible computer lines, serving different segments of the market. Plans were created for an 8000 series of computers, but a few visionaries within the company argued that creating yet another computer with its own new instruction set would only increase the confusion and escalate manufacturing and support costs. IBM engineers did not even plan to make the different models within the 8000 series computers compatible with each other. Such developments showed that IBM lacked a long-range vision.

An IBM electrical engineer turned manager, Robert O. Evans, led the charge to create a New Product Line that would cancel the 8000 series and completely replace all the computer systems that IBM manufactured with a uniform architecture of new machines. Frederick Phillips Brooks Jr., who earned a doctorate in applied mathematics from Harvard University in 1956, was the systems planning manager for the 8000 series and fought against Evans's plans. After the corporation decided to go with the plan proposed by Evans, the canny engineer asked Brooks to become a chief designer of the new system. Gene M. Amdahl (1922–), another brilliant young engineer, joined Brooks in designing the System/360.

Honeywell cemented the need for the System/360 computers when its Honeywell 200 computers, introduced in 1963, included a software utility allowing the new Honeywell computer to run programs written for the IBM 1401 computer. The 1401 was a major source of IBM profits, and the cheaper Honeywell 200 threatened to sweep the low-end market for data processing machines.

When Amdahl and Brooks decided that they could no longer work with each other, Evans solved the problem by keeping Amdahl as the main system designer and moving Brooks over to head the difficult task of creating a new operating system for the System/360. Initially, the designers of the operating system chose to create four different operating systems, for different sizes of machines, to be labeled I, II, III, and IV. This plan, which was based on Roman numerals and did not include compatibility between the different systems, was canceled in early 1964 because it conflicted with the overall design goal of system compatibility. The resulting OS/360

The IBM System/360 mainframe computer with peripheral devices. IBM Corporate Archives.

proved to be a difficult challenge, and even after the System/360 was announced in April 1964, the operating system was too full of bugs to be released. Part of the reason that the operating system fell behind is that IBM did not charge for software and thus thought of itself primarily as a hardware vendor, not a software vendor. But the OS/360 experience showed that software was becoming more important, and IBM executives paid more attention to software efforts thereafter. In the decade from 1960 to 1969, the fraction of total research and development efforts at IBM devoted to software rose from one-twentieth to one-third.

Brooks had wanted to retire and go to work at a university, but remained another year to help work the bugs out of OS/360. His experiences with this project led him to write *The Mythical Man-Month: Essays on Software Engineering* in 1975, which became a classic treatise in the field. A man-month is how much work a person can do in a month. If a task is said to take twenty man-months, then one person must work twenty months, or ten people must work two months. As the OS/360 project fell behind, IBM added more programmers to the project, which bloated the size of the teams, making communications between team members more complex and actually increasing the difficulty of completing the project. Brooks compared large programming projects to falling into a tar pit, and pointed out that programming should be a disciplined activity similar to engineering, with good process control, teamwork, and adherence to good design principles. Brooks also warned against the second system effect, where programmers who disciplined themselves on their first project relaxed and got intellectually lazy on their second project.

In 1965, after spending half a billion dollars on research and another $5 billion on development, IBM shipped the first System/360 machine to a customer. Within a year, a deluge of orders forced IBM to dramatically expand their manufacturing facilities. By the end of 1966, a thousand System/360 systems were being built and sold every month. The company found that its gamble had paid off, and the company increased its workforce by 50 percent in the next three years to keep up with demand, reaching almost a quarter of a million employees. By the end of the decade, IBM dominated at least 70 percent of the worldwide computer market.

The System/360 achieved its goal of upward and downward compatibility, allowing programs written on one system to run on a larger or smaller system. Now standardized peripheral devices, such as printers, disk drives, and terminals, would work on any of the System/360 machines. By having more uniform equipment, IBM also reined in manufacturing costs. IBM had earlier used the same strategy to dominate the market for punched-card machines, making a uniform family of machines that came in different models.

The IBM engineers played it safe with the technology in the System/360, choosing to use solid logic technology (SLT) instead of integrated circuits. SLT used transistors in a ceramic substrate, a new technology that could be mass produced more quickly. Though IBM advertised the System/360 as a third-generation computer, the technology remained clearly second generation. The System/360 standardized on 8 bits to a word, making the 8-bit byte universal. The System/360 also provided the features necessary to succeed as both a business data processing computer and a number-crunching scientific computer. IBM priced their systems as base systems, then added peripherals and additional features at an extra cost. IBM did such a good job of creating standardized computers that some computers were built with additional features already installed, such as a floating-point unit, and shipped to customers with those features turned off by electronic switches. Some customers, especially graduate students at universities, turned on the additional features to take advantage of more than the university paid for.

In the interests of getting their project done faster, the OS/360 programmers chose not to include dynamic address translation, which allowed programs to be moved around in memory and formed an important foundation of time-sharing systems. IBM fixed this problem and some of the other technical problems in the System/360 series with its System/370 series, introduced in 1970, by adding dynamic address translation, which became known as virtual memory.

The IBM System/360 became so dominant in the industry that other computer manufacturers, such as RCA with its Spectra 70 series, created their own System/360-compatible machines, competing with IBM in their own market with better service and cheaper prices. The British ICL 2900 series was System/360 compatible, as was the Riad computer series built behind the Iron Curtain for Soviet and Eastern European use.

After his instrumental role in designing the IBM 704, IBM 709, and System/360, Gene M. Amdahl grew frustrated that IBM would not build even more powerful machines. IBM based its customer prices proportional to computer processing power, and more powerful computers proved too expensive if IBM retained that pricing model. Amdahl retired from IBM in 1970 and founded his own company, Amdahl Corporation, to successfully build IBM-compatible processors that cost the same but were more powerful and took up less space that comparable IBM machines. Amdahl made clones of IBM mainframes a decade before clones of IBM personal computers completely changed the personal computer market.

BIRTH OF THE SOFTWARE INDUSTRY

By the mid-1960s, a small but thriving software services industry existed, performing contracts for customers. One of these companies, Applied Data Research (ADR), founded in 1959 by seven programmers from Sperry Rand, was approached in 1964 by the computer manufacturer RCA to write a program to automatically create flowcharts of a program. Flowcharts are visual representations of the flow of control logic in a program and are very useful to designing and understanding a program. Many programmers drew flowcharts by hand when they first designed a program, but as the program changed over time, these flowcharts were rarely updated and became less useful as the changed program no longer resembled the original. After writing a flowcharting program, ADR asked RCA to pay $25,000 for the program. RCA declined on the offer, so ADR decided to call the program Autoflow and went to the hundred or so customers of the RCA 501 computer to directly sell the program to them. This was a revolutionary step and resulted in only two customers, who each paid $2,400.

ADR did not give up. Realizing that the RCA market share was too small, the company rewrote Autoflow to run on IBM 1401 computers, the most prevalent computer at the time. The most common programming language on the IBM 1401 was called Autocoder, and Autoflow was designed to analyze Autocoder programming code and create a flowchart. This second version of Autoflow required that the programmer insert one-digit markers in their code indicating the type of instruction for each line of code. This limitation was merely an inconvenience if the programmer was writing a new program, but it was a serious impediment if the programmer had to go through old code adding the markers. Customers who sought to buy Autoflow wanted the product to produce flowcharts for old code, not new code, so ADR went back to create yet another version of Autoflow.

This third try found success. Now IBM 1401 customers were interested, though sales were constrained by the culture that IBM had created. Because IBM completely dominated the market and bundled their software and services as part of their hardware prices, independent software providers could not compete with free software from IBM, so they had to find market niches where IBM did not provide software. In the past, if enough customers asked, IBM always wrote a new program to meet that need and gave it away for free. Why should an IBM 1401 customer buy Autoflow when IBM would surely create the same kind of program for free? In fact, IBM already had a flowcharting program called Flowcharter, but it required the programmer to create a separate set of codes to run with Flowcharter, and did not examine the actual programming code itself.

Autoflow was clearly a superior product, but executives at ADR recognized the difficulty of competing against free software, so they patented their Autoflow program to prevent IBM from copying it. This led to the first patent issued on software in 1968, a landmark in the history of computer software. A software patent is literally the patenting of an idea that can only be expressed as bits in a computer, not as a physical device, as patents had been in the past.

ADR executives also realized that the company had a second problem. Computer programmers were used to sharing computer code with each other, freely exchanging magnetic tapes and stacks of punched cards. This made sense when software was free and had no legal value, but ADR could not make a profit if their customers turned around and gave Workflow away to their friends. Because there was no technical way to protect Autoflow from being copied, ADR turned to a legal agreement. Customers signed a three-year lease agreement, acquiring Autoflow like a piece of equipment, which they could use for three years before the lease must be renewed. With the success of the IBM System/360, ADR rewrote Autoflow again to run on the new computer platform. By 1970, several thousand customers used Autoflow, making it the first packaged software product. This success inspired other companies.

The next software product began as a file management system in a small software development company owned by Hughes Dynamics. Three versions, Mark I, Mark II, and Mark III, became increasingly more sophisticated during the early 1960s, running on IBM 1401 computers. In 1964, Hughes Dynamics decided to get out of the software business, but they had customers who used the Mark series of software and did not want to acquire a bad reputation by just abandoning those customers. John Postley, the manager who had championed the Mark software, found another company to take over the software. Hughes paid a software services firm called Informatics $38,000 to take their unwanted programmers and software responsibilities.

Postley encouraged Informatics to create a new version, Mark IV, that would run on the new IBM System/360 computers. He estimated that he needed half a million dollars to develop the program. With a scant $2 million in annual revenue, Informatics could not finance such a program, so Postley found five customers willing to put up $100,000 each to pay for developing Mark IV. In 1967, the program was released, selling for $30,000 a copy. More customers were found and within a year over $1 million of sales had been recorded, bypassing the success of the Autoflow product.

Informatics chose to lease their software, but for perpetuity, rather than a fixed number of years as ADR had chosen to do with Autoflow. This allowed Informatics to collect the entire lease amount up front, rather than over the life of the lease, as ADR did. This revision of the leasing model became the standard for the emerging industry of packaged software products. Informatics initially decided to continue to provide upgrades of new features and bug fixes to their customers free of charge, but that changed after four years and they began to charge for improvements and fixes to their own program, again setting the standard that the software industry followed after that.

Despite these small stirrings of a software industry, the computer industry was still about selling computer hardware. When the federal government looked at the computer industry, their antitrust lawyers found an industry dominated by one company to the detriment of effective competition. Under pressure from an impending antitrust lawsuit to be filed by the federal government, IBM decided in 1969 to unbundle its software and services from its hardware and sell them separately, beginning on January 1, 1970. This change created the opportunity for a vigorous community of software and service providers to emerge in the 1970s, competing directly with IBM. Even though IBM planned to unbundle their products, the federal government did file its antitrust lawsuit on the final day of the Johnson presidential administration (in January 1969), and the lawsuit lasted for thirteen years, a continual irritant distracting IBM management throughout that time.

Eventually the lawsuit disappeared, as it became apparent in the mid-1980s that IBM was on the decline and no longer posed a monopolistic threat.

An example of the effect of the IBM unbundling decision can be seen in the example of software for insurance companies. In 1962, IBM brought out their Consolidated Functions Ordinary (CFO) software suite for their IBM 1401 computer, which handled billing and accounting for the insurance industry. Because large insurance companies created their own software, designing exactly what they needed, the CFO suite was aimed at smaller and medium-sized companies. Since application software was given away for free until 1970, other companies who wished to compete with IBM had to create application software also. Honeywell competed in serving the life insurance industry with their Total Information Processing (TIP) System, closely modeled on IBM's CFO software. With the advent of the System/360, IBM brought out their Advanced Life Information System (ALIS) and gave it away to interested customers, though it was never as popular as the CFO product. After unbundling, literally dozens of software companies sprang up, offering insurance software. By 1972, 275 available applications were listed in a software catalog put out by an insurance industry association. Despite the example of insurance software applications, software contractors still dominated the emerging software industry, with $650 million in revenue in 1970, as opposed to $70 million in revenue for packaged software products in that same year.

Another type of computer services provider also emerged in the 1960s. In 1962, an IBM salesman, H. Ross Perot (1930–), founded Electronic Data Systems (EDS) in Dallas, Texas. The company bought unused time on corporate computers to run the data processing jobs for other companies. Not until 1965 did EDS buy its first computer, a low-end IBM 1401. EDS grew by developing the concept of what became known as outsourcing, which is performing the data processing functions for other companies or organizations. In the late 1960s, the new Great Society programs of Medicare and Medicaid required large amounts of records processing by individual states, and EDS grew quickly by contracting with Texas and other states to perform those functions. Further insurance, social security, and other government contracts followed, and by the end of the decade the stock value of EDS had passed $1 billion.

BASIC AND STRUCTURED PROGRAMMING

Even with the new second-generation programming languages, such as FORTRAN and COBOL, programming remained the domain of mathematically and technically inclined people. At Dartmouth College a pair of faculty members and their undergraduate students aimed to change that by developing a system and language for other Dartmouth students to use who were not majors in science or engineering. The Dartmouth team decided on an ambitious project to build both an interactive time-sharing operating system based on using teletype terminals and a new, easy-to-use programming language. In 1964, a federal grant allowed Dartmouth to purchase a discounted GE-225 computer. Even before the computer arrived, General Electric arranged for the Dartmouth team to get time on other GE-225 computers to create their Beginner's All-purpose Symbolic Instruction Code (BASIC) system. Dartmouth faculty taught BASIC in only two mathematics classes, second-term calculus and finite mathematics, where students were allowed to use an open lab and learn programming.

Clearly based on FORTRAN and ALGOL, BASIC used simple keywords, such as PRINT, NEXT, GOTO, READ, and IF THEN. General Electric adopted BASIC as their commercial time-sharing system, and within several years BASIC was ported to computers from other manufacturers. BASIC became the most widely known programming language because of its ease of use and because personal computers in the 1970s and 1980s adopted BASIC as their entry-level language. Early forms of the language were compiled, though the personal computer implementations were usually interpreted. Compiled programs are programs that have been run through a compiler to convert the original source code into binary machine code ready to be executed in the central processing unit. Interpreted code is converted to machine code one line at a time as the program is run, resulting in much slower execution. Compiled programs only have to be compiled once, while interpreted programs have to be interpreted every time that they run, resulting in a waste of computing resources.

All the early programming languages used some form of "goto" statements to unconditionally transfer control from one section of the program to another section. This method led to what became known as "spaghetti code," which is how a programmer felt when trying to follow the overlapping paths of logic in a program. This problem particularly occurred when programs were modified again and again, with ever more layers of logical paths intertwined with earlier logical paths. Programmers recognized that this was a serious problem, but did not know what to do about it.

The Dutch computer scientist Edsger W. Dijkstra (1930–) came to the rescue. The son of a chemist and mathematician, Dijkstra almost starved to death during the famine in the Netherlands at the end of World War II. After obtaining doctorates in theoretical physics and computer science, Dijkstra made a name for himself in the 1950s and 1960s as an innovative creator of algorithms, developing the famous shortest-path algorithm and the shortest spanning tree algorithm. He also contributed work on the development of mutual exclusion to help processes work together in multiprogramming systems. In 1968, as an eminent programmer, he sent an article to the *Communications of the ACM* journal, "A Case against the Goto Statement." The editor of the journal, Niklaus Wirth (1934–), chose to publish the article as a letter to the editor in order to bypass the peer-review process in the journal and speed up its publication. Wirth also picked a more provocative title: "The Goto Statement Considered Harmful."

Dijkstra showed that the goto statement was actually unnecessary in higher-level languages. Programs could be written without using the goto and thus be easier to understand. This insight led to "structured programming," and newer languages, such as C and Pascal (the latter designed by Wirth), allowed the goto to only act within the scope of a function or procedure, thus removing the bad effects of the instruction. Structured programming, the dominant programming paradigm in the 1970s and 1980s, allowed programmers to build larger and more complex systems that exhibited fewer bugs and were easier to maintain. Structured programming is only useful in the higher-level languages, since on the level of machine code, the actual bits that run on a CPU, the goto instruction, called a jump instruction, is still necessary and pervasive.

SUPERCOMPUTERS

Seymour Cray (1925–1996) showed his passion for electronics as a child, building an automatic telegraph machine at the age of only ten in a basement laboratory that his indulgent parents equipped for him. After service in the Army during World War II as a radio operator, Cray earned a bachelor's degree in electrical engineering and master's degree in applied mathematics before entering the new computer industry in 1951. He worked for Engineering Research Associates (ERA), designing electronic devices and computers. When ERA was purchased by Remington Rand (later called Sperry Rand), Cray designed the successful UNIVAC 1103 computer.

In 1957 a friend left Sperry Rand to form Control Data Corporation (CDC). Cray followed him and was allowed to pursue his dream of building a powerful computer for scientific computing. The result was the Control Data 1604 in 1960, built for the U.S. Navy. The most powerful computer in the world at that time, it was built entirely of transistors. The new category of supercomputer had been born, successors to the IBM Stretch project and the Sperry Rand LARC projects of the late 1950s. Cray continued to design new supercomputers, and the Control Data 6600, released in 1964, included a record 350,000 transistors. Supercomputers were used on the most difficult computing problems, such as modeling weather systems or designing complex electronic systems. Annoyed at the dominance of CDC in the new supercomputer field, IBM engaged in questionable business practices, such as announcing future supercomputer products that they did not ship, that led CDC to file an antitrust suit in 1968. The suit was settled in CDC's favor in 1973.

In 1972, Cray left CDC to found his own company, Cray Research, in his hometown of Chippewa Falls, Wisconsin. CDC generously contributed partial funding to help the new company. Cray was famous for his intense focus and hard work, though he played hard also; besides sports, he enjoyed digging tunnels by hand on his Wisconsin property.

In 1976, the Cray-1 was released, costing $8.8 million, with the first model installed at Los Alamos National Laboratory. Using vector processing, the Cray-1 could perform thirty-two calculations simultaneously. A refrigeration system using Freon dissipated the intense heat generated by the closely packed integrated circuits. Other improved systems followed, the Cray X-MP in 1982, the Cray-2 in 1985, and the Cray Y-MP in 1988. The last machine was the first supercomputer to achieve over a gigaflop in speed (1 billion floating-point operations per second); by contrast, the Control Data 6600 in 1964 could only do a single megaflop (1 million floating-point operations per second). Every Cray machine pushed the technology envelope, running at ever faster clock speeds and finding new ways of making more than one processor run together in parallel. The name Cray was synonymous with supercomputers, though the company's share in the supercomputing market fell in the 1990s as parallel-processing computers from other companies competed to build ever more powerful supercomputers. In early 1996, Cray Research merged with Silicon Graphics, Incorporated (SGI), and Cray died as a result of injuries from an automobile accident later that year.

MICROPROCESSORS

In 1968, Robert Noyce and Gordon E. Moore decided to leave Fairchild Semiconductor to found Intel Corporation. The two founders of Fairchild Semiconductor raised $500,000 of their own money and obtained another $2,500,000 in commitments from venture capitalists on the basis on their reputations and a single-page proposal letter. Intel had a product available within a year, a 64-bit static random access memory (RAM) microchip to replace magnetic core memory. IBM had already created such a technology and used it in its mainframe computers for temporary storage, but did not sell it separately. The Intel microchip crammed about 450 transistors onto the chip.

In 1970 Intel also introduced dynamic random access memory technology, which required a regular electric refreshing on the order of 1,000 times a second to keep the bit values stable. Magnetic core memories retained their bit values even if the power was turned off, while the new Intel technologies lost everything when power was cut. After only a couple of years, computer system designers adapted to this change because the new memory chips were so much cheaper and faster. Intel also licensed their technology to other microchip manufacturers so that they were not the sole source of the memory chips, knowing that computer manufacturers felt more comfortable having multiple suppliers.

Intel also invented erasable programmable read-only memory (EPROM) microchips in 1970. EPROMs are ROM chips with a window on top. By shining ultraviolet light into the window, the data on the microchip are erased and new data can be written to the microchip. This technology served the controller industry well, making it easy to embed new programs into their controllers. The EPROM provided a significant portion of Intel's profits until 1984. In that year, the market for microchips crashed; within nine months, the price of an EPROM dropped by 90 percent. Japanese manufacturers had invested heavily in the memory chip market and markets for other kinds of microchips, and manufacturing overcapacity drove prices below any conceivable profit margins. American memory chip manufacturers filed a legal suit alleging illegal dumping by the Japanese. The federal government became involved and while most American memory chip manufacturers withdrew from the market, the EPROM market was saved. By the mid-1980s, Intel no longer needed the EPROM market because they were then chiefly known as a manufacturer for their fourth major invention: the microprocessor.

In April 1969, Busicom, a Japanese manufacturer of calculators, approached Intel to manufacture a dozen microchips that they had designed for a new electronic calculator. Ted Hoff (1937–), who earned a doctorate in electrical engineering from Stanford University in 1962, was assigned to work with Busicom. Hoff determined that the Japanese design could be consolidated into just five chips. Intel convinced the Japanese engineers to allow them to continue trying to make even more improvements. The Japanese agreed and Hoff finally got the count down to three: a read-only memory (ROM) microchip, a random access memory (RAM) chip, and a microprocessor. The 4-bit microprocessor, called the Intel 4004, contained all the central logic necessary for a computer on a single chip, using about 2,000 transistors on the chip. Stanley Mazor (1941–) helped with programming the microprocessor, and Federico Faggin (1941–) did the actual work in silicon.

By March 1971, the microprocessor had been born. Intel executives recognized the value of the invention, but Busicom had negotiated an agreement giving them the rights to the new microchip. When Busicom began to experience financial difficulties, they wanted to negotiate a lower price for the microprocessors. Intel agreed to this lower price as long as Busicom allowed Intel to pay back the $65,000 in research money that Busicom had originally paid to Intel in return for Intel gaining the right to sell the microprocessor to other companies. Busicom agreed, and Intel offered the microprocessor for sale.

While the 4004 was still in development, Hoff designed another microprocessor, the 8-bit Intel 8008. This chip was again developed for an outside company, Computer Terminals Corporation (CTC). When CTC

The Intel 4004 microprocessor. Copyright Intel Corporation.

could not buy the microprocessor because of financial difficulties, Intel again turned to selling it to other customers. The Intel 8008 found a role as an embedded data controller and in dedicated word-processing computers. The Intel 8008 led to the 8-bit Intel 8080, brought to market in 1974, which became the basis of the first personal computer. The fourth generation of computer hardware, which is still ongoing, is based on microprocessors and ever more sophisticated integrated circuits. Intel and other companies sold 75 million microprocessors worldwide in just 1979, a strong indication of the outstanding success of Hoff's invention less than a decade later.

By 1960, less than 7,000 electronic digital computers had been built worldwide. By 1970, the number of installed electronic digital computer systems stood at about 130,000 machines. Yet computers remained expensive, found only in workplace or research settings, not in the home. In the 1970s the microprocessor became the key technology that enabled the computer to shrink to fit the home.

5

Personal Computers: Bringing the Computer into the Home

THE ALTAIR 8800

When Ted Hoff (1937–) of Intel created the Intel 4004 microprocessor, a complete central processing unit (CPU), the potential to build a small computer—a microcomputer—existed. Intel management wanted to stay out of end-user products sold directly to the end customer, so they did not take the next obvious step and create the first microcomputer. The rapid release of Intel's 8008 and 8080 microprocessors soon led a programmer, Gary Kildall (1942–1994), to begin creating a rudimentary operating system for the Intel microprocessors. Kildall and other computer hobbyists had a dream to create a "desktop" computer—a singular computer for their own personal use.

Electronic hobbyists were part of a small community of experimenters who read magazines like *Popular Electronics* and *Radio Electronics*, attended conventions and club meetings devoted to electronics, and built home electronic systems. They often shared their discoveries with each other. The technical director of *Popular Electronics*, Les Solomon, liked to spur the development of electronic technology by asking for contributions about a particular topic. The submissions that met Solomon's standards would get published in the magazine. In 1973 Solomon put out a call for "the first desktop computer kit." A number of designs were submitted, but all fell short, until Edward Roberts contacted Solomon and the cover story of the January 1975 issue introduced the new Altair 8800.

Edward Roberts (1941–) was born in Miami, Florida. From an early age he had two primary, seemingly disparate, interests in life: electronics and medicine. Family obligations and financial constraints caused him to pursue electronics. At the time of the Les Solomon challenge, Roberts ran his Micro Instrumentation and Telemetry Systems (MITS) calculator company, one of the first handheld calculator companies in the United States, in Albuquerque, New Mexico. Small companies like his were running into serious competition in the calculator market from big players like Texas Instruments and Hewlett-Packard. Roberts decided that he would devote his resources to try to meet Solomon's challenge and build a "desktop" computer with the hope of selling it to hobbyists. He realized that this was a big gamble because no one knew what the market for such a machine might be. He designed and developed the Altair for over a year before he sent a description to Solomon.

The name for the computer came about when Roberts wondered aloud what he should call the machine and his daughter suggested the Altair, since that was the name of the planet that the starship *Enterprise* was visiting that night on *Star Trek*. Oddly enough, science fiction writers and moviemakers before about 1970 did not foresee the rise of personal computers in the homes of average people. Perhaps large, monolithic computers made better villains in stories.

The Altair 8800 microcomputer was based on the 8-bit Intel 8080 microprocessor and contained only 256 bytes of memory. The kit cost $397 and came completely unassembled. A person could pay $100 more if they wanted to receive the Altair 8800 already assembled. The microcomputer had no peripherals: no keyboard, computer monitor, disk drive, printer, software, operating system, or any input or output device other than the toggle switches and lights on the front panel of the machine. Programs and data were loaded into memory through the toggle switches, using binary values, and the results of a program run were displayed as binary values on the lights. The Altair was a true general purpose computer, a von Neumann machine, with the capacity for input and output, even if rudimentary.

Roberts knew that peripheral devices would have to come later. To accommodate integrating them into the machine, the Altair had an open bus architecture. It consisted of a motherboard that held the CPU, and expansion slots for cards (circuit boards) that connected to a computer monitor or television, disk drives, or printers. Communication between the CPU and the expansion slots occurred through a bus, an electronic roadway by which the CPU checks to see what device on the computer needs attention.

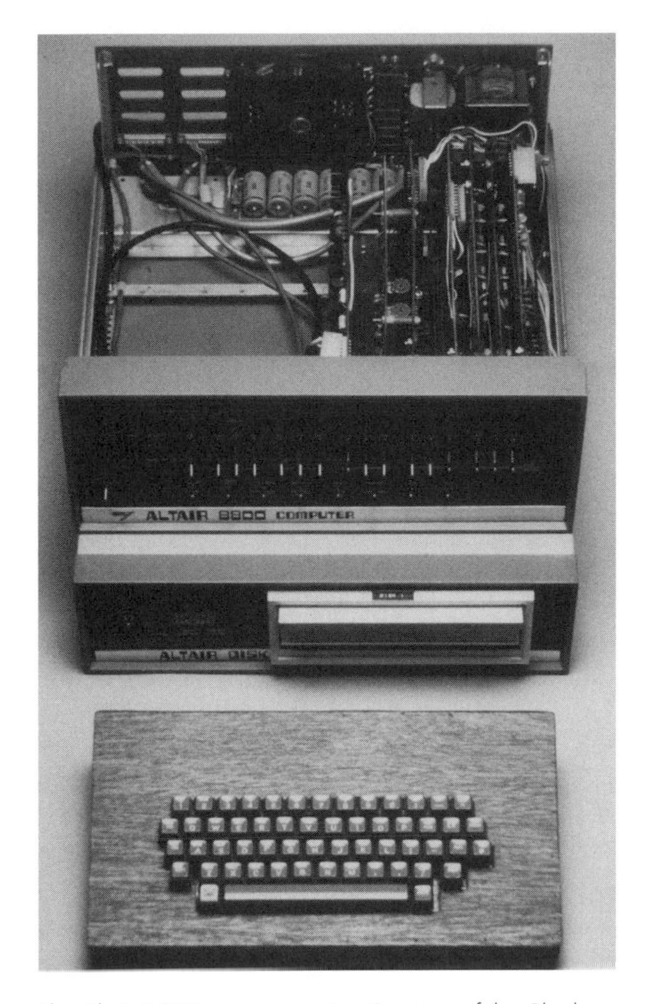

The Altair 8800 microcomputer. Courtesy of the Charles Babbage Institute, University of Minnesota, Minneapolis.

Four thousand orders for the Altair came to MITS within three months of publication of the *Popular Electronics* article describing the machine, demonstrating a surprisingly large market for home computers. Roberts had trouble obtaining parts, found that parts were not always reliable, and was unprepared to quickly manufacture that many machines, so it took months for MITS to fulfill the orders.

Despite the problems, electronic hobbyists were willing to purchase the microcomputer and put up with long delivery times and other problems. Altair clubs and organizations quickly sprang into existence to address the potential and needs of the machine. Some of these hobbyists became third-party manufacturers and created many of the peripherals needed by the machine, such as memory boards, video cards that could be attached to a television, and tape drives for secondary storage.

Although often given credit for inventing the personal computer, Roberts did not create the first inexpensive "desktop" computer. In France, Andre Thi Truong created and sold a microcomputer called the Micral based on the Intel 8008 in 1973. Truong sold 500 units of his Micral in France, but the design never was published in the United States. Though the Altair was not first, the size of the electronic hobbyist market in the United States and the open nature of the Altair's design contributed to the speedy development of microcomputers in the United States. All later development of microcomputers sprang from the Altair 8800, not the Micral.

ORIGINS OF MICROSOFT

Microsoft was started by Paul Allen (1953–) and Bill Gates (1955–), and owed its origins to the success of the Altair 8800. Allen was born to librarian parents who inspired his many interests. Gates was born to William Henry Gates Jr., a prominent attorney, and Mary Maxwell, a school teacher turned housewife and philanthropist.

Allen and Gates grew up together in Washington State. They were both enthusiastic about computing technology, and Gates learned to program at age thirteen. The enterprising teenagers both worked as programmers for several companies, including automotive parts supplier TRW, without pay, just for the fun of it. While in high school they created a computer-like device that could measure automotive traffic volume. They called the company Traf-O-Data. The company was short-lived but useful for the two in gaining business experience. Gates may have also created one of the first worms—a program that replicates itself across systems—when he created a program that moved across a network while he was still a junior in high school.

Gates was a student at Harvard University and Allen was working for Honeywell Corporation in Boston when Roberts's *Popular Electronics* article was published. Allen called Roberts in Albuquerque and found MITS had no software for the machine, so Allen called Gates and they decided to get involved. The two young men were so confident in their technical abilities, and believed that they could draw on the simple BASIC compiler they had already created for the Traf-O-Data machine, that they told Roberts they had a BASIC programming language for the Altair already working. Six weeks later they demonstrated a limited BASIC interpreter on the

Altair 8800 to Roberts. Roberts was sold on the idea and licensed the interpreter from Allen and Gate's newly formed company Micro-Soft (they later dropped the hyphen). Roberts also hired Allen as his one and only programmer, with the official title of director of software. Gates dropped out of Harvard to help improve the interpreter and build other software for MITS. The BASIC interpreter made operation of the Altair so much easier, opening up the machine to those who did not want to work in esoteric Intel microprocessor machine code.

MORE MICROCOMPUTERS

Altair was shipping microcomputers out to customers as fast as they could make them, and by the end of 1976, other companies began creating and selling microcomputers as well. A company called IMSAI used the Intel 8080 to create its own microcomputer and soon competed with MITS for market leadership. IMSAI gained some Hollywood fame by appearing in the movie *WarGames* as the microcomputer used by the main character played by Matthew Broderick. Companies like Southwest Technical Products and Sphere both used the more powerful Motorola 6800 microprocessor to create their own machines. The company Cromemco developed a computer around the Zilog Z80 chip, a chip designed by former Intel engineer Federico Faggin (1941–). MOS Technology, a semiconductor company, created a microcomputer around their own 6502 microprocessor, then sold the technology to Commodore and later to Atari. Radio Shack began to look for a machine that it could brand and sell in their stores.

Roberts had not patented the idea of the microcomputer, nor did he patent the idea of the mechanism through which the computer communicated with its components: the bus. Hobbyists and newly formed companies directly copied the Altair bus, standardized it so that hardware peripherals and expansion cards might be compatible between machines, and named it the S-100 bus. This meant that engineers could create peripherals and expansion cards for microcomputers that might work in more than just the Altair.

It became obvious to Roberts that the competition was heating up not just for computers but also for the peripherals on his own machine. Most of the profit came from peripherals and expansion cards, so Roberts tried to secure his position by requiring that resellers of the Altair 8800 only sell peripherals and expansion cards from MITS. Most refused to follow his instructions. Manufacturing problems continued as well, and to protect their sales of a problem-plagued 4K memory expansion card, MITS linked purchase of the card to the popular Micro-Soft BASIC. BASIC normally cost $500, but cost only $150 if purchased with a MITS memory card. This strategy did not work because a large number of hobbyists simply began making illegal copies of the software and bought memory cards from other manufacturers or made their own.

Seeking a new direction, MITS gambled on the future and released a new Altair based around the Motorola 6800. Unfortunately, hardware and software incompatibility between the new machine and the older 8800 machine, as well as the limited resources MITS had to assign to supporting both machines, did not help MITS in the market. In December 1977, Roberts sold MITS to the Pertec Corporation, and the manufacture of Altairs ended a year later. Roberts left the electronics industry and became a medical doctor, able to afford his long-time dream based on the profits from selling MITS. He later went on to combine electronics and medicine, creating a suite of medical laboratory programs in the mid-1990s.

Despite the demise of MITS and the Altair, the microcomputer revolution started by that machine had just began. Some fifty different companies developed and marketed their own home microcomputers. Many companies would quickly see their own demise. Others were successful for years to come. Commodore introduced their PET in 1977 and followed with even easier-to-use and cheaper models, the VIC-20 and Commodore 64, both based on the MOS 6502 microprocessor. Atari introduced their 400 and 800 machines, also based on the 6502 microprocessor. Radio Shack began to sell their TRS-80 (referred to in slang as the TRASH-80) in their stores nationally in 1977, helping to introduce computing to nonhobbyists.

THE APPLE II

The genesis of the Apple computer is found in Homestead High School in Sunnyvale, California, where students were often children of the numerous computer engineers and programmers who lived and worked in the area. Many of these children showed interest in electronic technology, including Steve Wozniak (1950–), one of the future founders of Apple computer, and often known just by his nickname: "Woz." One of Wozniak's first electronic devices simulated the ticking of a bomb. He placed it in a friend's school locker as a practical joke. The principal of the school found the device before the friend and suspended Woz for just two days in those more lenient times.

By 1971, Woz had graduated and was working a summer job between his first and second year of college, when he began to build a computer with an old high school friend, Bill Fernandez. They called it the Cream Soda Computer because of the late nights they spent building it and drinking the beverage. By Woz's account, the machine worked, but when they tried to show it to a local newspaper reporter, a faulty power supply caused it to burn up. This story shows how hobbyists were working to create the microcomputer, since the Cream Soda Computer's inauspicious debut came two years before the debut of the Altair 8800. Fernandez also introduced Woz to Steve Jobs (1955–), the other future cofounder of Apple Computer.

Jobs was another Silicon Valley student and, by most accounts, a bright, enterprising, and persuasive young man. He once called William Hewlett (1913–), one of the founders of Hewlett-Packard, and convinced Hewlett to lend him spare electronics parts. Jobs was twelve years old at the time. Though Jobs was five years younger than Woz when they met, they shared a common affection for practical jokes, and the two got on well. One of their first enterprises together proved rather dubious. They constructed "blue boxes," an illegal device that allows an individual to make free phone calls, and sold them to their friends. Jobs also obtained a summer job at Atari, a video game company newly founded in 1971 by Nolan Bushnell (1943–). Jobs enlisted Woz to help him program a game that Bushnell had proposed, even though Woz was already working full time at Hewlett-Packard. The game, Breakout, became an arcade hit.

Woz began working on another microcomputer in 1975. It was not a commercial product and was never intended to be, just a single circuit board in an open wooden box. Jobs, however, saw commercial potential and convinced Woz that it had a future. They called it the Apple I. As pranksters fond of practical jokes, they decided to begin the company on April Fool's Day in 1976. The price of the machine was $666.66.

Jobs's ambition went beyond the handmade Apple I and after consulting with Bushnell, he decided to seek venture capital. He was introduced to Mike Markkula (1942–), a former marketing manager at Fairchild and Intel who had turned venture capitalist. Markkula became convinced that the company could succeed. He secured $300,000 in funding from his own sources and a line of credit.

The Apple II, designed by Woz and based on the MOS 6502 microprocessor, was introduced in 1977. The Apple II cost $790 with 4 kilobytes of RAM, or $1,795 with 48 kilobytes of RAM. The company made a profit by the end of the year as production doubled every three months. Though Apple was hiring and bringing in money, Woz remained working full time at Hewlett-Packard, requiring Jobs to turn his arts of persuasion on Woz to convince him to come work at Apple full time.

The Apple II came in a plastic case that contained the power supply and keyboard. It had color graphics and its operating system included a BASIC interpreter that Woz had written. With a simple adapter, the Apple II hooked up to a television screen as its monitor. The Apple II was an attractive and relatively reliable machine. Many elementary and secondary schools purchased the Apple II across America, making it the first computer that many students came in contact with. The microcomputer's open design allowed third-party hardware manufacturers to build peripherals and expansion cards. For example, one expansion card allowed the Apple II to display eighty columns of both upper and lower case characters, instead of the original forty columns of only uppercase characters. Programming the Apple II was fairly simple, and many third-party software products were created for it as well. This ease of programming allowed the machine to reach broad acceptance, with the most popular programs on the hobbyist microcomputers being games like MicroChess, Breakout, and Adventure. Microcomputers, however, were not thought of as business machines. A program named VisiCalc changed that.

Steve Wozniak and the Apple II microcomputer. Copyright Roger Ressmeyer/CORBIS.

VisiCalc (short for visible calculations) was the first spreadsheet program. A spreadsheet is a simple table of cells in columns and rows. The columns and rows go beyond the boundaries of the screen and can be scrolled to either up and down or left and right. Cells contain text, numbers, or equations that can summarize and calculate values based on the contents of other cells. The spreadsheet emulates a paper accounting sheet but is far more powerful because it can change the value cells dynamically as other cells are modified. The idea had been around on paper since the 1930s as a financial analysis tool, but the computer made it a truly powerful idea.

Dan Bricklin (1951–) and Bob Frankston (1949–), two Harvard MBA students, wrote VisiCalc on an Apple II in Frankston's attic. They released their program in October 1979 and were selling 500 copies a month by the end of the year. A little more than a year later, VisiCalc was selling 12,000 copies a month at $150 per copy. Other powerful business programs were introduced as well. For example, John Draper, a former hacker known as Cap'n Crunch, wrote EasyWriter, the first word-processing application for the Apple II. Compared to all other programs, however, VisiCalc was so successful in that it drove people to purchase the Apple II just to run it. A new term described this kind of marketing wonder software: the "killer app." A killer app (or killer application) is a program that substantially increases the popularity of the hardware it runs on. Apple continued to prosper, and in 1981 the company had sales of $300 million a year and employed 1,500 people.

THE IBM PC

On August 12, 1981, a new player joined the ranks of microcomputer manufacturers: Big Blue. IBM saw the possibility of using microcomputers on business desks, then decided they needed to get on the ground with their own microcomputer, and to do it quickly. Their intention was to dominate the microcomputer market the same way they dominated the mainframe marketplace, though they anticipated that the microcomputer market would remain much smaller than the mainframe market.

In 1980, IBM approached the problem of going to market with a microcomputer differently than they had for any other hardware they had produced. They chose not to build their own chipset for the machine like they had for their mainframes and minicomputers, and so the new microcomputer used the 16-bit Intel 8088, a chip used in many other microcomputers. IBM also learned from the successes of the Altair 8800 and other microcomputer pioneers by recognizing that IBM needed the many talents of the microcomputer world to build the peripherals and software for their PC. They also decided to go outside IBM for the software for the machine, including the operating system. To facilitate third-party programming and hardware construction, IBM did a few other things that never would have occurred in the mature market of mainframes. IBM created robust and approachable documentation and an open bus-type architecture similar to that of the Altair's S-100 bus. Recognizing the change in the market landscape, IBM also sold the machine through retail outlets instead of only through their established commercial sales force.

Searching for applications for its microcomputer, IBM contacted Microsoft and arranged a meeting with Gates and his new business manager, Steve Ballmer (1956–), in Microsoft's Seattle area offices. Gates's mother may have played a role in Microsoft's eventual overwhelming success, since she sat on the board of the United Way with a major executive at IBM and he recognized Microsoft as her son's business. Gates and Ballmer put off a meeting with Atari to meet with IBM. Atari was in the process of introducing computers for the home market based around the MOS 6502 microprocessor. For the meeting with IBM, Gates and Ballmer decided to look as serious as possible and put on suits and ties—a first for them in the microcomputer business. In another first, they signed a confidentiality agreement so that both Microsoft and IBM would be protected in future

The IBM PC. IBM Corporate Archives.

development. Microsoft expressed interest in providing software applications for the new machine.

IBM also needed an operating system and went to meet with Gary Kildall at Digital Research Incorporated (DRI). Kildall had written an operating system called Control Program for Microcomputers (CP/M) that worked on most 8-bit microprocessors, as long as they had 16 kilobytes of memory and a floppy disk drive. This popular operating system ran on the IMSAI and other Altair-like computers, and by 1981 sat on over 200,000 machines with possibly thousands of different hardware configurations. Before CP/M, the closest thing to an operating system on the microcomputer had been Microsoft BASIC and Apple's BASIC. CP/M was much more powerful and could work with any application designed for the machines. However, IBM hesitated at paying $10 for each copy of CP/M. IBM wanted to buy the operating system outright at $250,000. Talking again with Gates, they became convinced that they might be better off with a whole new operating system, because CP/M was an 8-bit operating system and the 8088 was 16-bit CPU. So, despite Microsoft not actually owning an operating system at the time, IBM chose Microsoft to provide their microcomputer operating system. Microsoft was a small company among small software companies, bringing in only $8 million in revenue in 1980, when VisiCalc brought in $40 million in revenue in the same year.

Microsoft purchased a reverse-engineered version of CP/M from Seattle Computer Products called SCP-DOS, which they reworked into the Microsoft Disk Operating System (MS-DOS), which IBM called the Personal Computer Disk Operating System (PC-DOS), to run on the Intel 8088 microprocessor. CP/M and MS-DOS not only shared the same commands for the user, but also even the internal system calls for programmers were the same. Kildall considered a lawsuit at this brazen example of intellectual property theft, but instead reached an agreement with IBM for the large company to offer his operating system as well as the Microsoft version. Unfortunately, when the product came out, IBM offered PC-DOS at $40 and CP/—86 at $240. Not many buyers went for the more expensive operating system.

A mantra had existed in the computer world for many years: no one ever got fired for buying IBM. With IBM now in the microcomputer market, businesses that would never think about buying a microcomputer prior to the IBM PC were in the market for them. With the introduction of the IBM PC, microcomputers were now referred to generically as personal computers or PCs, and were suddenly a lot more respectable than they had been. Another killer application appeared that also drove this perception. A program called Lotus 1-2-3, based on the same spreadsheet principle as

VisiCalc, pushed the PC. The introduction of the program included a huge marketing blitz with full-page ads in the *Wall Street Journal*. To the investors of Wall Street and executives in larger corporations, the software and the hardware manufacturer were legitimate.

For a couple years in the early 1980s, it was not clear where the market would go. Osborne Computer, founded in 1981, created the first real portable personal computer. The Osborne 1 used a scrollable 5-inch screen; contained a Zilog Z80 microprocessor, 64 kilobytes of RAM, and two floppy disk drives; and was designed to fit under an airplane seat. The portable ran CP/M, BASIC, the WordStar word-processing software, and the Supercalc spreadsheet. The Osborne 1 sold for only $1,795, and soon customers were buying 10,000 units a month. Other portable personal computers, almost identical to the Osborne, quickly followed from Kaypro and other manufacturers. In 1980 and 1981, other large computer manufacturers began to bring out personal computers, such as Hewlett-Packard with its HP-85, Xerox with its Star, and Digital Equipment Corporation (DEC) with its Rainbow (a dual-processor machine that could run both 8-bit and 16-bit software).

Despite the other efforts, the successful combination of IBM and Microsoft killed most of the rest of the personal computer market. By the end of 1983, IMSAI was gone, Osborne had declared bankruptcy, and most of the 300 computer companies that had sprung up to create microcomputers that were not compatible with IBM PCs had mostly disappeared. Kildall's DRI also began its downward spiral as CP/M became less important. By 1983, it looked like there would soon be only two companies left selling microcomputers on a large scale to battle for supremacy: IBM and Apple. *Time* magazine also noticed the importance of the microcomputer when they chose the PC as their Man of the Year for 1982, the only time that they chose a machine for an honor that usually went to important international leaders.

XEROX PARC, THE GUI, AND THE MACINTOSH

With the introduction of the IBM PC and Microsoft DOS, Apple faced serious competition for the first time, and Jobs turned to formulating a response. As an operating system, DOS adequately controlled the machine's facilities, but few would call the user interface intuitive. Users typed in cryptic commands at the command line in order to get the machine to do anything. Jobs visited the Xerox Palo Alto Research Center (PARC) in 1979 and came away with a whole different idea for a user interface.

Established in 1970, Xerox PARC was initially headed by Robert Taylor (1932–), previously director of the Information Processing Techniques Office at the Advanced Research Projects Agency (ARPA) in the Pentagon. Taylor helped lay the groundwork for the network that became the Internet, and brought his skill at putting together talented people and resources to PARC. Scientists and engineers at PARC quickly established themselves as being on the cutting edge of computing science. In 1973, PARC created a computer they called the Alto that used a bitmapped graphical display, a graphical user interface (GUI), a mouse, and programs based on the "what you see is what you get" (WYSIWYG) principle. The GUI used two- or three-dimensional graphics and pointing mechanisms to the graphics, with menus and icons, as opposed to the old method of using text commands at a operating system command prompt. Jobs also saw an Ethernet network linking the computers on different engineers' desks to each other and to laser printers for printing sharp graphics and text.

Though structured programming only truly started to be established in the industry in the 1970s, programmers at PARC needed more, so they created Smalltalk, an object-oriented programming language (OOP) better suited for writing a GUI and other graphical programs. Variants of the Simula languages designed at the Norwegian Computing Center in Oslo, Norway, in the 1960s were the first examples of OOP, but Smalltalk came to be considered the purest expression of the idea. Structured programs usually separated the data to be processed from the programming code that did the actual processing. Object-oriented programs combined data and programming code into objects, making it easier to create objects that could be reused in other programs. Object-oriented programming required thinking in a different paradigm than structured programmers did, and OOP did not gain widespread acceptance until the late 1980s.

With this plethora of riches, practically every major innovation that would drive the computer industry for the next decade, generating hundreds of billions of dollars in revenue, Xerox remained a copier company in its heart. Xerox introduced the 8010 "Star" Information System in 1981, a commercial version of the Alto, but priced it so high at $40,000 that a system and peripherals cost about as much as a minicomputer. Though about 2,000 Star systems were built and sold, this was a failure compared to what Jobs eventually did with the concepts. Because of the failure of Xerox to exploit their innovations, scientists and engineers began to leave PARC to found their own successful companies or find other opportunities. Jobs eventually convinced many of the engineers at Xerox PARC to come over to Apple Computer.

Many of Xerox's innovations implemented the prior ideas of Douglas C. Engelbart (1925–), a visionary inventor. Raised on a farm, Englelbart entered Oregon State College in 1942, majoring in electrical engineering under a military deferment program during World War II. After two years, the military ended the deferment program because of a more immediate need for combat personnel versus a longer term need for engineers. Engelbart elected to join the Navy and became a technician, learning about radios, radar, sonar, teletypes, and other electronic equipment. He missed the fighting and returned to college in 1946. Two years later he graduated and went to work for the National Advisory Committee for Aeronautics (NACA), a precursor to the National Aeronautics and Space Administration (NASA). He married in 1951, and feeling dissatisfied with his work at NACA, he sought a new direction in his life.

After considerable study he realized that the amount of information was growing so fast that people needed a way to organize and cope with the flood, and computers were the answer. Engelbart was also inspired by the seminal 1945 article, "As We May Think," by the electrical engineer Vannevar Bush (1890–1974), who directed the American Office of Scientific Research and Development during World War II. Bush had organized the creation of a mechanical differential analyzer before the war and after the war envisioned the use of computers to organize information in a linked manner that we now recognize as an early vision of hypertext. Electronic computers in 1951 were in their infancy, with only a few dozen in existence. Engelbart entered the University of California, Berkeley, and earned a master's degree in 1952 and a Ph.D. in electrical engineering in 1955, with a speciality in computers.

Engelbart became an employee at the Stanford Research Institute (SRI) in 1957 and his paper, "Augmenting Human Intellect: A Conceptual Framework," laid out many of the concepts in human-computer interaction he had been working on. He formed his own laboratory at SRI in 1963, called the Augmentation Research Center (ARC). Engelbart's team of engineers and psychologists worked through the 1960s on realizing his dream, the NLS (oNLine System). Engelbart wanted to do more than automate previous tasks like typing or clerical work; he wanted to use the computer to fundamentally alter the way that people think. In a demonstration of the NLS at the Fall Joint Computer Conference in December 1968, Engelbart showed the audience onscreen video conferencing with another person back at SRI, 30 miles away; an early form of hypertext; the use of windows on the screen; mixed graphics-text files; structured document files; and the first mouse. This influential technology demonstration became known as "the mother of all demos."

While often just credited with inventing the mouse, Engelbart had also developed the basic concepts of groupware and networked computing.

Engelbart's innovations were ahead of their time, requiring expensive equipment that retarded his ability to innovate. A computer of Engelbart's at SRI became the second computer to join the ARPAnet in 1969, an obvious expansion of his emphasis on networking. ARPAnet later evolved into the Internet. In the early 1970s, several members of Engelbart's team left to join the newly created PARC, where ample funding led to rapid further development of Engelbart's ideas. It only remained for Steve Jobs and Apple to bring the work of Engelbart and PARC to commercial fruition.

Steve Jobs has said of his Apple I, "We didn't do three years of research and come up with this concept. What we did was follow our own instincts and construct a computer that was what we wanted." Job's next foray into computer development used the same approach. The first attempt by Apple to create a microcomputer that used the GUI interface was the Lisa, based on a 16-bit Motorola 68000 microprocessor, released in 1983. The Lisa was expensive and noncompatible with both the Apple II computer line and the rest of the DOS-oriented microcomputer market, and did not sell well.

After Jobs become disenchanted with the Lisa team during production, he decided to create a small "skunk works" team to produce a similar but less expensive machine. Pushed by Jobs, the team built a computer and small screen combination in a tan box together with keyboard and mouse: the Apple Macintosh. The Macintosh, also based on the Motorola 68000 microprocessor, was the first successful mass-produced GUI computer.

The Macintosh's public unveiling was dramatic. During the 1984 Super Bowl television broadcast, a commercial flickered on that showed people clothed in grey trudging like zombies into a large bleak auditorium. In the front of the auditorium, a huge television displayed a talking head similar to the character "Big Brother" from George Orwell's novel *1984*, droning on. An athletic and colorfully clothed woman chased by characters that look like security forces runs into the room. She swings a sledgehammer into the television. The television explodes, blowing a strong dusty wind at the seated people. A message comes on the screen: "On January 24th, 1984 Apple Computer will introduce Macintosh. And you'll see why 1984 won't be like 1984." The commercial reference to Orwell's novel, where Big Brother is an almost omnipotent authoritarian power, is intriguing. Although never stated, it was not hard to guess that Apple was likening to Big Brother to Apple's nemesis, IBM.

The Macintosh (or "Mac," as it was affectionately called) quickly garnered a lot of attention. Sales were initially stymied by hardware limitations, since the Mac had no hard drive and limited memory, and lacked extensive software. Eventually, Apple overcame these initial limitations, allowing the machine to fulfill its promise. Even with the initial problems, the Macintosh suddenly changed the competitive landscape. The development of Aldus Pagemaker by Paul Brainard (1947–) in 1985, the first desktop publishing program, became the killer application for the Macintosh, making it a successful commercial product, just as VisiCalc had made microcomputers into useful business tools.

Engelbart's contributions were lost in popular memory for a time, even at Apple, which at one point claimed in a famous 1980s lawsuit against Microsoft to have effectively invented the GUI. Yet, by the mid-1980s, people in the computer industry began to take notice of Engelbart's contributions and the awards began to flow. Among his numerous awards were a lifetime achievement award in 1986 from PC Magazine; a 1990 ACM Software System Award; the 1993 IEEE Pioneer award; the 1997 Lemelson-MIT Prize, with its $500,000 stipend; and the National Medal of Technology in 2000.

Jobs was forced out of Apple in 1985 by John Scully (1939–), the man he had handpicked to be the new head of Apple. Jobs reacted by founding NeXT Inc., intending to build the next generation of personal computers. The NeXTcube experienced considerable problems, but finally came to market in 1990, being built in a completely automated factory. Designed around a 32-bit Motorola 68030 microprocessor, containing 8 megabytes of memory, and including a 256-megabyte magneto-optical drive for secondary storage instead of a floppy disk drive, the NeXTcube ran a sophisticated variant of the UNIX operating system and included many tools for object-oriented programming. The NeXTcube impressed technical people because of the sophistication of its software, but that software ran slowly, the computer cost $9,999, and apparently the window for introducing a completely new microcomputer architecture had passed for a time. Only some 50,000 units were sold, and the company lost money.

Jobs also cofounded Pixar Animation Studios after purchasing the computer graphics division of Lucasfilm, made famous by the *Star Wars* movies. Pixar created computer-animated movies, using proprietary software technology that they developed, and by concentrating on their storylines, not deadlines, began a string of successes with the full-length feature film, *Toy Story*, in 1995. In 1996, the board of directors of his old company, Apple, after suffering business losses, asked Jobs to return to head the company. He did so on the condition that Apple would buy NeXT Inc. They did so, and the NeXT operating system and programming tools were integrated into the Apple Macintosh line. Jobs successfully turned Apple around, relishing a sense of vindication.

IBM PC CLONES

When IBM first approached Microsoft, Bill Gates successfully convinced IBM that their PC should follow the direction of open architecture that they had begun in their hardware by having their PC be able to support any operating system. He pointed to the success of VisiCalc, where software drove hardware sales. Gates figured that he could compete successfully with any other operating system. In many ways, this was not a large gamble. Gates understood that a paradigm-shifting operating system might come along to supplant DOS, but he also knew from his experiences in the microcomputer world that users tended to stick to a system once they had acquired experience in it. This is known as technological lock-in.

Gates also argued to IBM that since Microsoft was at risk for potentially having its operating system on the PC replaced by a competitor's, Microsoft should be free to sell its operating system to other hardware manufacturers. IBM bought the argument and opened the door for clones. Gates was acutely aware of the experience of the Altair with its open architecture, which quickly led to clones. The open architecture of the PC meant that third parties could also clone the IBM PC's hardware.

While Apple kept its eye on the feared giant of IBM, other companies grabbed market share from both Apple and IBM by creating IBM PC clones. IBM PCs could be cloned for three reasons: because Intel could sell their microprocessors to other companies, not just IBM; Microsoft could also sell their operating system to the clone makers; and the Read-Only Memory Basic Input/Output System (ROM BIOS) chips that IBM developed could be reverse engineered. ROM BIOS chips are memory microchips containing the basic programming code to communicate with peripheral devices, like the keyboard, display screen, and disk drives. A clone market for Apple computer did not emerge because Apple kept a tight legal hold on their Macintosh ROM BIOS chips, which could not easily be reverse engineered.

One of the first clone makers was also one of the most successful. Compaq Computer was founded in 1982 and quickly produced a portable computer that was also an IBM PC clone. When the company began to sell their portable computers, their first year set a business record when they sold 53,000 computers for $111 million in revenue. Compaq moved on to building desktop IBM PC clones and continued to set business records. By 1988, Compaq was selling more than $1 billion of computers a year. The efforts of Compaq, Dell, Gateway, Toshiba, and other clone makers continually drove down IBM's market share during the 1980s and 1990s. The clone makers produced cheaper microcomputers with more power and features than the less nimble IBM could.

By 1987, the majority of personal computers sold were based on Intel or Intel-like microchips. Apple Macintoshes retreated into a fractional market share, firmly entrenched in the graphics and publishing industries, while personal computers lines from Atari and Commodore faded away. The success of the clone makers meant that the terms "personal computer" and "PC" eventually came to mean a microcomputer using an Intel microprocessor and a Microsoft operating system, not just an IBM personal computer.

Intel saw the advantages of the personal computer market and continued to push the microprocessor along the path of Moore's Law. The 8088 was a hybrid 8/16-bit microprocessor with about 29,000 transistors. The 16-bit Intel 80286 microprocessor, introduced in 1982, had 130,000 transistors. The 32-bit Intel 80386 microprocessor, introduced in 1985, had 275,000 transistors. The 32-bit Intel Pentium microprocessor, introduced in 1993, contained 3.1 million transistors. The Pentium Pro, introduced in 1995, contained 5.5 million transistors; the Pentium II, 7.5 million transistors; and the Pentium III, released in 1999, 9.5. million transistors. The Pentium IV, introduced in 2000, used a different technological approach and reached 42 million transistors on a single microchip.

One of the major reasons for the success of the Intel-based personal computer is that other companies also made Intel-like chips, forcing Intel to continually strive to improve their products and keep their prices competitive. Without this price pressure, personal computers would have certainly remained more expensive. In the early 1980s, at the urging of IBM, Intel licensed their microprocessor designs to other computer chip manufacturers, so that IBM might have a second source to buy microprocessors from if Intel factories could not keep up with demand. In the 1990s, after Intel moved away from licensing their products, only one competitor, Advanced Micro Devices (AMD), continued to keep the marketplace competitive. AMD did this by moving from just licensing Intel technology to reverse-engineering Intel microprocessors and creating their own versions. By the early 2000s, AMD was designing different features into the microprocessors than were found in comparable Intel microprocessors and still remaining mostly compatible.

SOFTWARE INDUSTRY

After IBM's unbundling decision in 1969, which led to the sale of software and hardware separately, the software industry grew rapidly. In 1970, total sales of software by U.S software firms was less than half a billion dollars. By

1980, U.S. software sales reached $2 billion. Most of these sales in the 1970s were in the minicomputer and mainframe computer markets. Sales of software for personal computers completely revolutionized the software industry, dramatically driving up sales during the 1980s. In 1982, total sales of software in the U.S. reached $10 billion; in 1985, $25 billion. The United States dominated the new software industry, which thrived in a rough-and-tumble entrepreneurial atmosphere.

Creators of personal computer software did not come from the older software industry, but sprang out of the hobbyist and computer games communities and sold their software like consumer electronics products, in retail stores and through hobbyist magazines, not as a capital product with salespeople in suits visiting companies. Hobbyists and gamers also demanded software that was easier to learn and easier to use for their personal computers than the business software that was found on mainframes. This emphasis on human factors design became an important part of the software industry and eventually even affected how mainframe business software was designed.

In about 1982, as the IBM PC and its clones became dominant in the marketplace, the software market became more difficult for the young hobbyist to enter. VisiCalc contained about 10,000 lines of programming code, something that a pair of programmers could easily manage, whereas Lotus 1-2-3, the product that pushed VisiCalc out of the market, contained about 400,000 lines of code, which required a team effort. VisiCalc had sold about 700,000 copies since its launch, when Lotus 1-2-3, propelled by $2.5 million in advertising, sold 850,000 copies in its first 18 months.

As the cost of entering the software marketplace went up, an interesting alternative marketing model emerged. Beginning in about 1983, programmers who created a useful program often offered it to other people as shareware. This usually meant that anyone who wanted to could use the program, and a donation was requested if the program proved useful. Among the more useful programs distributed under this scheme were a word processor, PC-WRITE; a database, EASY-FILE; and a modem control program, PC-TALK. Many minor games were also distributed as shareware.

GAMES

By 1982, annual sales in the United States of computer games stood at $1.2 billion. Computer games had their origin in mechanical pinball machines. The first electric pinball machine was built in 1933, and later electronics were included to make the machines more sophisticated and flashier. The first true computer game was invented by Massachusetts Institute of Technology (MIT) graduate student Steve Russell in 1962 on a Digital Equipment Corporation PDP-1. MIT, Stanford University, and the University of Utah were all pioneers in computer graphics and some of the few places in the early 1960s where a programmer could actually use a video terminal to interact with the computer. Russell's game, Spacewar, graphically simulated two spaceships maneuvering and firing rocket-propelled torpedoes at each other. Using toggle switches, the users could change both the speed and direction of their ships and fire their torpedoes. Other students added accurate stars for the background, and a sun with a gravity field that correctly influenced the motion of the spaceships. The students also constructed their own remote controllers so that their elbows did not grow tired from using the toggle switches on the PDP-1.

Nolan Bushnell (1943–), educated at the University of Utah, played Spacewar incessantly at the university, inspiring him to write his own computer games while in school. After graduating, Bushnell designed an arcade version of Spacewar, called Computer Space, and found a partner willing to manufacture 1,500 copies of the game for the same customers who purchased pinball machines, jukeboxes, and other coin-operated machines. Far too complex for amateurs to play, the game failed to sell.

Bushnell did not give up, but partnered with a fellow engineer to found a company called Atari in 1972. While Bushnell worked on creating a multiplayer version of Computer Space, he hired an engineer and assigned him to create a simple version of ping-pong that could be played on a television set. Pong became a successful arcade video game, then Bushnell in 1975 partnered with Sears to sell a version called Home Pong in their stores. Home Pong attached to the television set at home. The game sold wildly and Atari released its Atari 2600 in 1977, a home unit that could play many games that each came on a separate cartridge. Though Bushnell had been forced out of the company in 1978, his dream of a commercially successful version of Spacewar was realized in 1979, when Atari released Asteroids, which became their all-time best-selling game.

Other companies also competed in the home video market, but Atari defined the home video game market in the eyes of many, until the company took awful losses in 1983 as the market for game consoles crashed. Part of the reason for the crash was that personal computer games were becoming more popular. Nintendo revived the game console market in 1986, and both Sega and Sony, all Japanese companies, joined the competition. Nintendo had learned from the mistakes of Atari, and kept tight legal and technical control over the prices of game cartridges, so that excessive competition would not drive the prices of games down so far that profit disappeared. In the 1990s, game console systems and games for personal computers became so popular that the revenue in the game market surpassed the revenue generated by movies in Hollywood.

Games for personal computers existed from the start of the personal computer revolution, but did not become a powerful market force until about the time that game consoles stumbled in 1983. The personal computer, with a keyboard, provided a better interface for more sophisticated games, rather than just straight arcade-style games. Games such as Adventure from Adventure International (founded 1978), Zork from Infocom (founded 1979), Lode Runner from Broderbund (founded 1980), and Frogger from Sierra Online (founded 1980) defined the memories that many new personal computer users of that period have.

In the late seventies, games called multiuser dungeons (MUDs) appeared in Britain and the United States. The games were not created for commercial sale, but for fun, and ran on early networks and bulletin board systems. Players used a text interface, making their way through dungeons, fighting monsters, and interacting with other players.

In 1997, Ultima Online, a massively multiplayer online role-playing games (MMORPG), showed a new direction for gaming with the graphical power and sophistication of single-user personal computer games combined with the versatility and multiplayer challenge of MUDs. Later online games, such as Everquest and Runescape, also successfully followed Ultima Online. South Korea, because of its heavily urban population, had over 70 percent of all households connected to the Internet via high-speed broadband connections in the early 2000s. The online game Lineage in South Korea, released in 1998, became so popular that by 2003 nearly 2 million people played it every month, out of a total population of less than 49 million. Lineage is a medieval fantasy epic, which seemed to be the preferred format for successful online games.

MICROSOFT ASCENDENT

From its infant beginnings of offering BASIC on the Altair and other early microcomputers, Microsoft grew quickly as its executives effectively took advantage of the opportunities that the IBM PC offered. Microsoft actively aided the growth of the PC clone market, since every IBM PC and PC clone required an operating system, enabling Microsoft to earn revenue on every PC sold. Microsoft also created a single game, Flight Simulator, first released in 1983, that was so demanding of the PC's hardware and software that running Flight Simulator became a litmus test as to whether a new clone model was truly compatible enough with the PCs from IBM.

Eventually over 100 million copies of DOS were sold. Using the revenue from its dominant operating system, Microsoft developed further versions of DOS and funded the development of other software packages. The original DOS 1.0 contained only 4,000 lines of programming code. DOS 2.0, released in 1983, contained five times that much code, and DOS 3.0, released in 1984, doubled the amount of code again, reaching 40,000 lines. From early on, Microsoft developed well-regarded compilers and other programming tools. They also developed other types of application software, such as word processors and spreadsheets, but were not as successful in those product categories until the 1990s.

Microsoft saw the advantage of the GUI that Engelbart and PARC had invented, and they hired several top programmers from PARC. Microsoft developed early applications for the Apple Macintosh, even though Apple was a major competitor, and Microsoft also created their own GUI for DOS, called Windows. The first version shipped in 1985, and Microsoft soon followed that with a second version. Both versions were truly awful products: slow, aesthetically ugly, and mostly useless except for a few programs written to use them.

DOS was a primitive operating system at best, unable to effectively multitask or even effectively manage memory above a 640-kilobyte limit. IBM and Microsoft decided to jointly create OS/2, a next-generation operating system for the PC that would include multitasking, better memory management, and many of the features that minicomputer operating systems had. OS/2 1.0 was released in December 1987, and the second version, OS/2 1.10, released in October 1988, included a GUI called Presentation Manager. A severe shortage of memory microchips drove up the price of RAM memory from 1986 to 1989. In late 1988, a mere 1 megabyte of RAM cost about $900. OS/2 required substantially more memory than DOS, and the high costs of memory inhibited the widespread adoption of OS/2.

Even while working on OS/2 and Presentation Manager, Microsoft persisted in its own Windows efforts. Version 3.0, released in 1990, was an astounding commercial success, prompting pundits to argue that Microsoft took three tries to get their products right. Two factors contributed to the success of Windows: the memory shortage had ended and more users found it easier to buy the extra memory that Windows demanded, and programmers at companies that made software applications had already been forced by the Macintosh and OS/2's Presentation Manager to learn how to program GUI programs. This programming knowledge easily transferred to writing software for the more successful Microsoft Windows.

IBM was never able to regain any momentum for OS/2, though OS/2 had matured into a solid operating system. When Microsoft decided to continue its Windows development efforts to the detriment of its OS/2 development efforts, Microsoft and IBM decided to sever the close partnership that had been characteristic in 1980s. By this time, IBM was in deep disarray as they lost control of the PC market to the clone makers, found that PCs were becoming the dominant market segment in the computer industry, and saw the mainframe market began to contract. IBM actually began to lose money, and lost an astounding $8.1 billion on $62.7 billion in revenue in 1993. That year, IBM brought in a new chief executive officer from outside the company, Louis V. Gerstner Jr. (1942–), who managed to financially turn the company around through layoffs and refocusing the business on providing services. IBM remained the largest computer company, but never dominated the industry as it once had. In contrast to IBM's size, Microsoft passed $1 billion a year in revenue in 1990.

Microsoft Windows was so phenomenally successful that in 1992, Microsoft actually began running television commercials, something that the computer industry rarely did, despite the example of the 1984 Apple commercial. Television, as a mass medium, had not been used up to then because the market for personal computers had not been a mass consumer product. Now it was.

Windows 3.0 was not really a new operating system, just a user interface program that ran on top of DOS. Microsoft created a new operating system, Windows NT, that contained the multitasking features, security features, and memory management that had made OS/2, UNIX, and other minicomputer operating systems so useful. Windows NT 3.1 came out in 1993, but was not particularly successful until Windows NT 4.0 came out in 1996. Microsoft now had two Windows operating system lines, one for business users and servers, and one for home consumers. With Windows 95, where Microsoft chose to change from version numbers based on release numbers to those based on years, Microsoft updated the consumer version of Windows. Windows 95 was an important product because the ease of use and aesthetic appeal promised by the GUI paradigm, successfully achieved by Apple over a decade earlier, had finally been achieved by Microsoft.

Microsoft regularly produced new versions of their operating systems, adding features that demanded ever larger amounts of processor power, RAM, and disk drive space with each release. These increasing demands promoted the sales of ever more powerful PCs, making PCs effectively obsolete within only a few years of manufacture. The PC market in the 1990s effectively became dominated by what became known as the Wintel alliance, a combination of the words Window and Intel. With Windows XP, released in 2001, Microsoft finally managed to merge their consumer and business operating systems into a single release, after several earlier failed attempts.

In 1983, Microsoft released Microsoft Word, their word-processing software application, principally written by a veteran from PARC. At that time, products like MicroPro's WordStar and WordPerfect dominated the word-processing market. Microsoft released a version of Word for the Macintosh in 1984 and came to dominate in that market segment on the Macintosh, but Word did not threaten the success of other word-processing applications on IBM PCs and PC clones until Windows 3.0 gave Microsoft developers a jump on the competition.

In the 1990s, Microsoft utilized its position as sole supplier of operating systems to PCs to compete against software application companies. Leading software products like Lotus 1-2-3, Harvard Graphics, WordPerfect, and dBase began to lose market share after Windows 3.0 changed the PC market direction from the command-line DOS to the GUI Windows. Lotus 1-2-3 3.0 and WordPerfect 5.1 were the best-selling software packages in electronic spreadsheets and word processors respectively in 1991. Only three years earlier Lotus earned more in gross revenue than Microsoft, and in 1991 Lotus earned only slightly less gross revenue than Microsoft. By the year 2000, some version of Windows was on over 90 percent of the personal computers in the world, and in application software, Microsoft's Excel and Word programs had replaced most of the market share once enjoyed by Lotus 1-2-3 and WordPerfect.

Microsoft battled repeated complaints and lawsuits that they unfairly used their dominance in the operating systems market segment to then dominate other software market segments. These complaints were based on two assertions: first, that Microsoft created undocumented system calls that allowed its own applications to take special advantage of the Windows operating system. They had also done this in DOS. Second was that Microsoft set up special deals with personal computer manufacturers, such as Compaq, Dell, or Gateway, where Microsoft sold their operating systems at a steep discount if the computer manufacturers would also only sell the Microsoft applications software at the same time. These original equipment manufacture (OEM) deals encouraged consumers to turn from buying their business applications software from retail stores to buying them from their computer manufacturer. The market for retail stores offering computer software collapsed, and those stores mostly disappeared during the 1990s. The federal government twice sued Microsoft for antitrust violations on their software distribution and pricing practices, and both times found against Microsoft, but no effective legal counteraction was ever taken.

Microsoft also aggressively entered any market that they thought might overshadow their dominance of personal computer software by possibly making personal computers less important, launching a version of Windows for personal digital assistants, Windows CE, in the late 1990s, and a game console system called the Xbox in 2001. Both efforts yielded mixed success, since the measure of success for Microsoft was market dominance.

Having developed Hodgkin's disease, Allen left active participation in Microsoft in 1982. Allen is still one of the richest men in the world, and has sponsored and invested in many endeavors, including Steven Spielberg's Dreamworks studio, major sport teams, the Experience Music Project museum in Seattle devoted to his guitar idol Jimi Hendrix, and SpaceShipOne, the first commercially funded piloted vehicle to reach space. In 2004, Gates was the richest man in the world, worth more than $80 billion, though he had placed a substantial part of his fortune in a philanthropic trust.

In 2004, Microsoft announced that it estimated that there were 600 million Windows PCs around the world, and expected that number to pass 1 billion in just six more years. Microsoft revenues continued to set records. In 2000, Gates resigned as chief executive officer, naming himself chief software architect so that he could be more involved in the technical direction of the company and less distracted by its day-to-day management. By 2004, Microsoft employed over 50,000 people, and had a total annual revenue of over $35 billion, of which over $26 billion was gross profit. Microsoft's practice of not usually paying stock dividends meant that they had accumulated a cash reserve of $56 billion and zero debt. The little company that Gates and Allen had founded in 1975 had grown to become one of the most profitable on the planet and part of the Dow Jones 30 Industrials, the world's most commonly quoted stock indicator.

6

Connections: Networking Computers Together

THE COLD WAR

In 1957, the rocket engineers of the Soviet Union embarrassed the United States by launching Sputnik 1, the first artificial satellite. This event provoked a strong political and cultural reaction in the United States—funding for education, especially science and engineering, increased, and federal funds for research and development in science and technology also rose. A space race rapidly emerged. As a struggle of competing ideologies, the Cold War conflict between the superpowers depended as much on prestige as military power, and the United States wanted to regain its prestige as the preeminent scientific and technological power on the planet.

The Advanced Research Projects Agency (ARPA) was formed in 1958 in response to Sputnik and the emerging space race. Because ARPA was an agency of the Pentagon, its researchers were given a generous mandate to develop innovative technologies. Though ARPA scientists and engineers did conduct their own research, much of the effort came through funding research at universities and private corporations.

In 1962, a psychologist from the Massachusetts Institute of Technology's Lincoln Laboratory, J.C.R. Licklider (1915–1990), joined ARPA to take charge of the Information Processing Techniques Office (IPTO). Licklider's intense interest in cybernetics and "man-computer symbiosis" was driven by his belief that computers could significantly enhance the ability of humans to think and solve problems. Licklider created a social network of like-minded scientists and engineers and wrote a famous 1963 memorandum to these friends and colleagues, called "Memorandum for Members and Affiliates of the Intergalactic Computer Network," in which he described some of his ideas for time sharing and computer networking. His IPTO office funded research efforts in time sharing, graphics, artificial intelligence, and communications, laying the conceptual and technical groundwork for computer networking.

TELEPHONES

Networking already existed in the form of telegraphs and telephones. Samuel F. B. Morse (1791–1872) invented the telegraph in 1844, allowing communications over a copper wire via electrical impulses that operators sent as dots and dashes. Alexander Graham Bell (1847–1922) invented the telephone in 1877, using an analog electrical signal to send voice transmissions over wires. Teletype systems were first patented in 1904 and allowed an automatic typewriter to receive telegraph signals and print out the message without a human operator.

In the 1950s, the U.S. military wanted their new SAGE computers to communicate with remote terminals, so engineers developed a teletype to send an analog electrical signal to a distant computer. In 1958, researchers at Bell Telephone Laboratories took the next step and invented the modem, which stood for modulator-demodulator. Modems converted digital data from a computer to an analog signal to be transmitted across phone lines, then converted that signal back into digital bits for the receiving computer to understand. In 1962, the Bell 103, the first commercial modem, was introduced to the market by American Telephone and Telegraph (AT&T), the parent company of Bell Labs, running at 300 baud, which transmitted 300 bits per second. Modem speed steadily increased, eventually reaching 56K in the mid-1990s.

Each computer manufacturer tended to define their own character set for both letters and numbers, even changing them from model to model, forcing programmers to convert data when transferring their files from one computer to another. The American National Standards Institute (ANSI) defined the American Standard Code for Information Interchange (ASCII) in 1963. This meant that the binary sequence for the letter "A" would be the same on all computers. While the rest of the industry turned to ASCII, especially when networking and personal computers became more common in the 1970s, IBM maintained its own standard, Extended Binary-Coded Decimal Interchange Code (EBCDIC), for decades.

PACKET SWITCHING

In the early 1960s, the Polish-born electrical engineer Paul Baran (1926–), who worked for the RAND corporation, a think tank funded by the American military, faced a problem. Simulations of an attack with nuclear weapons by the Soviet Union showed that even minor damage to the long-distance phone system maintained by the telephone monopoly AT&T would cripple national communications. The telephone system that had developed during the twentieth century was based on analog transmissions over lines connected to switches. When a person made a long-distance telephone call, an actual electrical circuit was created via numerous switches in a scheme called "circuit switching."

Baran had considerable experience with computers, including working on the original UNIVAC, and appreciated the value of digital electronics over analog electronics. Baran devised a scheme of breaking signals into blocks of data to be reassembled after reaching their destination. These blocks of data traveled through a "distributed network" where each "node," or communication point, could independently decide which path the block of information took to the next node. This allowed data to automatically flow around potential blockages in the network, and to be reassembled into a complete message at the destination. Baran called his scheme "hot potato" routing, because each network node would toss the message to another node rather than hold onto it.

The Pentagon and AT&T were not interested in Baran's scheme of distributed communications because it required completely revamping the technology of the national telephone system. A British team under the direction of Donald Davies (1924–), at the British National Physical Laboratory (NPL), also independently developed a similar scheme to Baran's, which they called "packet switching." Davies and his team went further than Baran and actually implemented their ideas, and by 1970 had a local area network running at the NPL that used packet switching.

ARPANET

In 1966, Robert Taylor (1932–), then head of the IPTO, noted that in his terminal room at the Pentagon he needed three different computer terminals to connect to three different machines in different locations around the nation. Taylor also recognized that universities working with the IPTO needed more computing resources. Instead of the government buying machines for each university, why not share machines? Taylor revitalized Licklider's ideas, secured $1 million in funding, and hired Larry Roberts (1937–), a twenty-nine-year-old computer scientist, to direct the creation of ARPAnet.

In 1965, Roberts, while working at MIT's Lincoln Laboratory, had supervised an ARPA-funded pilot project to have two computers communicate over a long distance. Two computers, one in Boston and the other in Santa Monica, California, sent messages to each other over a set of leased Western Union telephone lines. The connection ran slowly and unreliably, but offered a direction for the future. ARPAnet was the next logical step. Roberts drew on the work of Baran and Davies to create a packet-switched networking scheme. While Baran was interested in a communications system that could continue to function during a nuclear war, ARPAnet was purely a research tool, not a command and control system as is often reported in contemporary media accounts.

Universities were reluctant to share their precious computing resources and concerned about the processing load of a network on their systems. Wesley Clark (1927–), computer lab director at Washington University of St. Louis, proposed an Interface Message Processor (IMP), a separate smaller computer for each main computer on the network that would handle the network communication.

A small consulting firm in Cambridge, Massachusetts, Bolt Beranek and Newman (BBN), got the contract to construct the needed IMPs in December 1968. They decided that the IMP would only handle the routing, not the transmitted data content. As an analogy, the IMP looked only at the addresses on the envelope, not at the letter inside. Faculty and graduate students at the host universities created host-to-host protocols and software to enable the computers to understand each other. Because the machines did not know how to talk to each other as peers, the researchers wrote programs that fooled the computers into thinking they were talking to preexisting dumb terminals.

ARPAnet began with the installation of the first 900-pound IMP, which cost about $100,000 to build, in the fall of 1969 at the University of California, Los Angeles (UCLA), followed by three more nodes at the Stanford Research Institute (SRI); the University of California, Santa Barbara; and the University of Utah. Fifty kilobit per second (kbps) communication lines connected each node to each other. The first message transmitted between UCLA and SRI was "L," "O," "G," the first three letters of "LOGIN," then the system crashed. Initial bugs were overcome, and ARPAnet added an extra node every month in 1970. BBN continued to run ARPAnet for the government, keeping the network running through round-the-clock monitoring at their network operations center.

With a network in place, ARPAnet scientists and engineers turned to using the network to get useful work done. Transferring files and remote login were obvious and useful applications. In 1971, file-transfer protocol (FTP) was developed. The protocol originally required a user to authenticate themselves with a username and password, but a system of using anonymous FTP later allowed any user to download those files that had been made available for everyone. Remote login was achieved through a variety of programs, though telnet, developed in 1971, eventually became the standard.

Also in 1971, Ray Tomlinson (1941–), an engineer at BBN supporting ARPAnet, found himself working on a program called CPYNET (for "copynet"), designed to transfer files between computers. He realized that CPYNET could be combined with SNDMSG (for "send message"), a program designed to send messages to a user on the same computer, and send messages from one computer to another. Tomlinson did so, and e-mail (electronic mail) was born. Tomlinson also developed the address format user@computer that used the @ symbol and later became ubiquitous. Electronic mail became what later pundits would call the "killer application" of ARPAnet, its most useful feature and its most commonly used application.

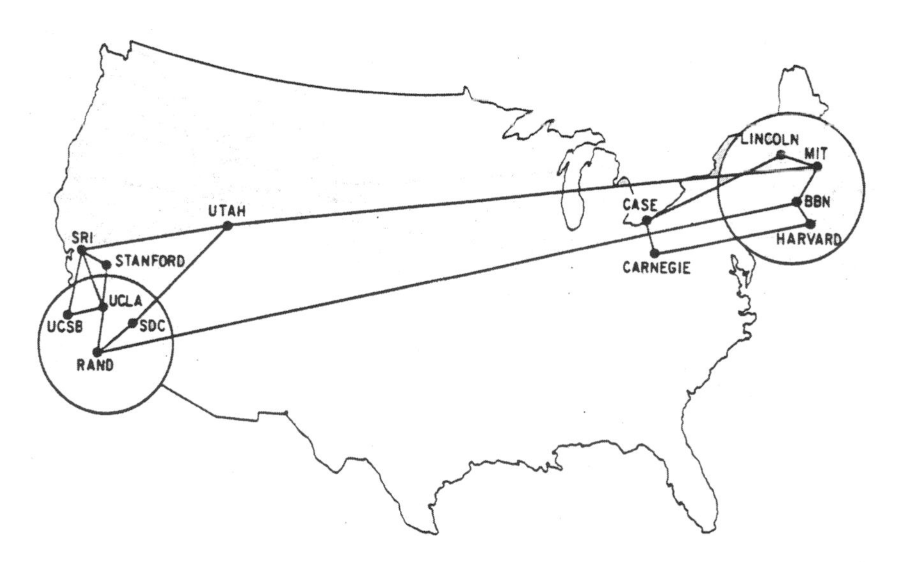

ARPAnet in December 1970. Courtesy of the Charles Babbage Institute, University of Minnesota, Minneapolis.

From the beginning of networking, programs had been designed to run across the network. As time when by, many of these programs used the same design structure, which became known as client-server systems. A server program provided some service, such as a file, or e-mail, or connection to a printer, while a client program communicated with the server program so that the user could use the service.

Roberts succeeded Taylor as head of the ITPO, and in 1972 Roberts arranged for a large live demonstration of ARPAnet at the International Conference on Computer Communications in Washington, D.C. None of the work on ARPAnet was classified, and the technical advances from the project were freely shared. The vision of what was possible with networking rapidly caught the imagination of scientists and engineers in the rest of the computer field. IBM announced their Systems Network Architecture (SNA) in 1974, which grew more complex and capable with each passing year. Digital Equipment Corporation released their DECnet in 1975, implementing their Digital Network Architecture (DNA). Other large computer manufacturers also created their own proprietary networking schemes.

THE BEGINNING OF WIRELESS NETWORKING

The Advanced Research Projects Agency also funded the effort by Norman Abramson (1932–) of the University of Hawaii to build AlohaNet in 1970. In addition to being one of the earliest packet-switching networks, AlohaNet broke new ground in two more ways. By transmitting radio signals between terminals through a satellite, AlohaNet became the first wireless network and first satellite-based computer network. One of the first technical hurdles to such a scheme was, how did the network program on a terminal know when it could send a radio signal? If two terminals sent a signal at the same time, the signals would interfere with each other, becoming garbled, and neither would be received by other terminals. The conventional answer was time-division multiple access (TDMA), where terminals coordinated their activity and only transmitted during their allocated time. For instance, perhaps each terminal would each get a fraction of a second and no two terminals could use the same fraction. The problem with this scheme was how to actually divide up the time slices and account for some terminals being used while others were offline TDMA tended to become more difficult to maintain as more terminals were added to the conversation.

AlohaNet had so many terminals that TDMA was impractical and a new scheme was developed, carrier sense multiple access with collision detection

(CDMA/CD). Under this scheme, any terminal could transmit whenever it wanted, but then listened to see if its transmission was garbled by another transmission. If the message went through, then everything was fine and the bandwidth was now free for any other terminal to use; if the signal became garbled, then the sending terminal recognized that it had failed and waited for a random amount of time before trying to send the same message again. This scheme, seemingly chaotic, worked well in practice as long as there were not too many terminals and as long as traffic was low enough so that there were not too many collisions. The scheme also allowed terminals to readily be added to and removed from the network without needing in any way to inform the other terminals about their existence.

Robert Metcalfe (1946–), a researcher at the innovative Xerox Palo Alto Research Center, visited Hawaii in 1972 and studied AlohaNet for his doctoral dissertation. Returning to Xerox, Metcalfe then developed Ethernet, using the CDMA/CD scheme running over local wire networks. Metcalfe left Xerox to cofound 3Com in 1979, a company that successfully made Ethernet the dominant networking standard on the hardware level in the 1980s and 1990s.

TCP/IP AND RFCs

ARPAnet originally used a set of technical communications rules called the network control protocol (NCP). NCP assumed that every main computer on the ARPAnet had identical IMP computers in front of them to take care of the networking. All the IMP machines were built by the same people, using the same designs, minimizing the risk of incompatibilities. This worked well, but NCP was not the only networking protocol available. Other companies developing their own networking schemes also developed their own set of proprietary protocols. Engineers at both BBN and the Xerox Palo Alto Research Center wanted to create a new set of network protocols that would easily enable different networks, each running their own set of unique protocols (such as NCP or SNA), to communicate with each other. This idea, called internetworking, would allow the creation of a network of networks.

Vint Cerf (1943–) is often called the "father of the Internet." As a graduate student, he worked on the first IMP at UCLA and served as a member for the first Network Working Group that designed the software for the ARPAnet. Bob Kahn (1938–) and Cerf first proposed Transport Control Protocol (TCP) in 1974 to solve the problem of internetworking, and Cerf drove the further development of protocols in the 1970s. The internetworking protocol eventually split into two parts: TCP and Internet Protocol (IP). TCP/IP was an open protocol, publicly available to everyone, with no restrictive patents or royalty fees attached to it, and ARPAnet switched to using TCP/IP in the late 1970s.

The philosophy behind Metcalfe's Ethernet heavily influenced TCP/IP. The NCP scheme had little error correction, because it expected the IMP machines to communicate reliably. TCP/IP could not make this assumption, and included the ability to verify that each packet had been transmitted correctly. In order for TCP/IP to work correctly, each machine must have a unique IP address, which comes in the form of four groups of numbers, for instance, 168.192.54.213. TCP/IP also made it simple to add and remove computers to the network, just as Ethernet could. In July 1977, an experiment with a TCP system successfully transmitted packets via the three types of physical networks that made up ARPAnet: radio, satellite, and ground connections. The packets began in a moving van in San Francisco, were transmitted via radio, crossed the Atlantic Ocean to Norway via satellite, bounced to London, then returned to the University of Southern California, a total of 94,000 miles in transit. This proof of concept became the norm as the ARPAnet matured.

The original team working on ARPAnet was called the Network Working Group, which evolved into the Internet Engineering Task Force (IETF) and the Internet Engineering Steering Group (IESG). These groups used the unique process of "requests for comments" (RFC) to facilitate and document their decisions. The first RFC was published in 1969. By 1989, with some 30,000 hosts connected to the Internet, 1,000 RFCs had been issued. Ten years later, millions of hosts used the Internet and over 3,000 RFCs had been reached. The RFC process created a foundation for sustaining the open architecture of the ARPAnet/Internet, where multiple layers of protocols provided different services. Jon Postel (1943–1998), a computer scientist with long hair and a long beard, edited the RFCs for almost thirty years before his death in 1998, a labor of love that provided a consistency to the evolution of the Internet. The actual work of the IETF is still performed in working groups, and anyone can join a working group and contribute their observations and work to the group, which will result in a new RFC.

INTERNET

In the 1960s, after introducing the modem, AT&T began to develop the technology for direct digital transmission of data, avoiding the need for modems and the inefficiency that came from converting to and from analog.

A lawsuit led to the Carterphone Decision in 1968, which allowed non-AT&T data communications equipment to be attached to AT&T phone lines, prompting other companies to develop non-AT&T modem and data communications equipment. In the 1970s, leased lines providing the digital transmission of data became available, including X.25 lines based on packet-switching technology. The availability of these digital lines laid the foundation for the further spread of wide area networks (WANs).

ARPAnet was not the only large network, only the first that paved the way. International Business Machines (IBM) funded the founding of Bitnet in 1984 as a way for large universities with IBM mainframes to network together. Within five years, almost 500 organizations had 3,000 nodes connected to Bitnet, yet only a few years later the network had disappeared into the growing Internet. The Listserv program first appeared on Bitnet to manage e-mail lists, allowing people to set up, in effect, private discussion groups. These e-mail lists could be either moderated or unmoderated. Unmoderated lists allowed anyone who wanted to join and send messages; moderated lists set up a person or persons as moderators, who controlled who could join the list and checked every e-mail that went through the list before passing them on to the general membership of the list. Moderated lists became more popular because they prevented a flood of superfluous e-mails from dominating the list and driving away members.

In 1981 the National Science Foundation (NSF) created the Computer Science Network (CSNET) to provide universities that did not have access to ARPAnet with their own network. In 1986, the NSF sponsored the NSFNET "backbone" to connect five supercomputing centers together. The backbone also connected ARPAnet and CSNET. The idea of the Internet, a network of networks, became firmly entrenched. The open technical architecture of the Internet allowed numerous innovations to easily be grafted onto the whole, and proprietary networking protocols were abandoned in the 1990s as everyone moved to using TCP/IP.

As mentioned before, TCP/IP only recognizes different computers' hosts by their unique IP number, such as 192.168.34.2. People are not very good at remembering arbitrary numbers, so a system of giving computers names quickly evolved. Each computer using TCP/IP had a file on it called "hosts" that contained entries matching the known names of other computers to their IP addresses. These files were each maintained individually, and the increasing number of computers connected to the ARPAnet/Internet created confusion. In 1983, a domain name system (DNS) was created, where DNS servers kept master lists matching computer names to IP addresses. A hierarchical naming system was also created, with computer names being attached to domain names and ending with the type of domain. Six extensions were created:

.com: commercial
.edu: educational
.net: network
.gov: government
.mil: military
.org: organization

When ARPAnet was dismantled in 1990, the Internet was thriving at universities and technology-oriented companies. In 1991, the federal government lifted the restriction on the use of the Internet for commercial use. The NSF backbone was later dismantled in 1995 when the NSF realized that commercial entities could keep the Internet running and growing on their own. The NSF backbone had cost only $30 million in federal money during its 9-year life, with donations and help from IBM and MCI (a telecommunications company). What began with four nodes in 1969 as a creation of the Cold War became a worldwide network of networks, forming a single whole. In early 2001, an estimated 120 million computers were connected to the Internet in every country of the world. As a global computer network interconnecting other computer networks, the Internet provided a means of communication unprecedented in human history.

BULLETIN BOARD SYSTEMS AND DIAL-UP PROVIDERS

In January 1978, when a severe snowstorm shut down the city of Chicago, two friends, Ward Christensen and Randy Suess, decided to develop a system to exchange messages. Christensen wrote the software and Suess put together the hardware, based on a homemade computer, using a S-100 bus and hand-soldered connections, running the CP/M operating system. They finished their effort in a month and called their system the Computer Bulletin Board Systems (CBBS), which allowed people to call in, post messages, and read messages. Modems at the time were rare, but the subsequent development of cheaper modems allowed computer hobbyists to set up their own Bulletin Board Systems (BBSs) and dial into other BBSs. Later enhancements allowed users to upload and download files, enter chat areas, or play games. Hundreds of thousands of BBSs eventually came and went, serving as a popular communications mechanism in the 1980s and early 1990s. A separate network connecting BBSs even emerged in the mid-1980s, called FidoNet, exchanging e-mail and discussion messages. At its height in 1995, FidoNet connected some 50,000 BBS nodes to each other. The coming of the public Internet in the 1990s doomed the BBS as a technology, though the social and special interest communities that had grown up around various BBSs transferred their communities to the Internet.

In 1969, CompuServe began as a time-sharing service in Columbus, Ohio. A decade later, in 1979, the company expanded to offer e-mail and simple services to home users of personal computers. In 1980, CompuServe offered the first real-time chat service with a program called CB Simulator that allowed users to simultaneously type in messages and have the results appear on each other's screens. From this humble beginning, what later became instant messaging was born. In the 1980s, CompuServe built its own countrywide network, which customers could use by dialing in with a modem to connect to large banks of modems that CompuServe maintained. CompuServe also offered the use of its network to corporations as a wide area network and expanded into Japan and Europe. CompuServe continually expanded the offerings that its customers paid to access, such as discussion groups, content from established national newspapers and magazines, stock quotes, and even a stock-trading service.

Sears and IBM created Prodigy, their own online service provider, in the 1980s, and it soon had over a million subscribers. In 1985, Steve Case (1958–), a computer enthusiast with a taste for business and filled with entrepreneurial zeal, founded Quantum Computer Services, a BBS for users of Commodore 64 personal computers. When Case wanted to expand and compete with the other online services, like CompuServe and Prodigy, he renamed his company America Online (AOL) in 1989.

In 1991, after the federal government lifted the restriction on the use of the Internet for commercial use, numerous Internet service providers (ISPs) sprang up immediately, offering access to the Internet for a monthly fee. In 1992, the Internet included a million hosts. CompuServe, AOL, and Prodigy began to provide access to the Internet to their customers, thus transforming these companies into instant ISPs. CompuServe became the largest ISP in Europe. Fueled by an aggressive marketing campaign, which included flooding the nation's mail with sign-up disks, AOL grew quickly. America Online (AOL) reached 1 million subscribers in August 1994, passed 2 million in February 1995, and peaked at 25 million subscribers in 2000. Over 1 million of those subscribers were in Germany, and AOL had over 5 million subscribers outside the United States in 2001. AOL grew so large that in 1997 they purchased CompuServe. Prodigy failed to successfully make the transition to being an ISP and faded away.

LOCAL AREA NETWORKS

The APRAnet and its successor, the Internet, were examples of WANs, where computers communicated across the street, across the nation, and even around the world. In the 1970s, research at the Xerox PARC, which had led to many innovations, also led to the creation of local area networks (LANs). A LAN is usually defined as a network for a room or a building. Ethernet provided one of the early standards for LAN computing, though other network card technologies, such as ARCNET, also appeared.

In the early 1980s, with easy new availability of large numbers of personal computers, various companies developed network operating systems (NOS), mainly to provide an easy way for users to share files and share printers. Later, LAN-based applications based on the NOS became available. The most successful of these network operating systems came from Novell. The company, founded in 1979 as Novell Data Systems, originally made computer hardware, but after the company was sold in 1983, the new president of the company, Raymond J. Noorda (1924–), turned them toward concentrating on software. That same year, the first version of NetWare came out. Novell ruled the LAN NOS market, achieving almost a 70 percent share and adding ever more sophisticated features to each version of NetWare. Novell created their own networking protocols, called IPX (Internet Packet eXchange) and SPX (Sequenced Packet eXchange), drawing on open networking standards that Xerox had published. Eventually, in the 1990s, as the Internet became ever more pervasive, Novell also turned to supporting TCP/IP as a basic protocol in NetWare. The dominance of NetWare rapidly declined in the late 1990s when Microsoft provided networking as a basic part of their Windows operating systems.

USENET

In 1979, graduate students at Duke University and the University of North Carolina wrote some simple programs to exchange messages between UNIX-based computers. This collection of programs, called news, allowed users to post messages to a newsgroup and read messages that other users had posted to that same newsgroup. The news program collected all the postings and then regularly exchanged them with other news programs via home-made 300-baud modems. The students brought their project to a 1980 Usenix conference. At that time, ARPAnet was only available to universities and research organizations who had defense-related contracts, so most universities were excluded from the network. Because so many universities had

UNIX machines, Usenix conferences allowed users to meet and exchange programs and enhancements to the UNIX operating system itself. The students proposed that a "poor man's ARPAnet" be created, called Usenet, based on distributing the news program and using modems to dial up other UNIX sites. To join Usenet, one had to just find the owner of a Usenet site who would allow you to download a daily news feed.

Usenet grew quickly, reaching 150 sites in 1981, 1,300 sites by 1985, and 11,000 sites in 1988. A protocol was eventually developed in the early 1980s, the network news transmission protocol (NNTP), so that news reader clients could connect to news servers. ARPAnet sites even joined Usenet because they liked the Usenet newsgroups. An entire culture and community grew up around Usenet, where people posted technical questions on many aspects of programming or computing and received answers within a day from other generous users. Usenet news discussion groups originally concentrated on technical issues, then expanded into other areas of interest. Anyone who wanted to could create a new newsgroup, though if it did not attract any postings, the newsgroup eventually expired. Programmers developed a way to encode binary pictures into ASCII, which could then be decoded at the other end, and picture newsgroups, including an enormous number of pornographic pictures, became a major part of the daily Usenet news feed.

As part of the culture of Usenet, a social standard of net etiquette evolved, eventually partially codified in 1995 in RFC1855, "Netiquette Guidelines." One such rule is that words in all-capital letters are the equivalent of shouting. A set of acronyms evolved also, such as IMO for "in my opinion" or LOL for "lots of laughs," as well as some symbols, such as :-) for a smile and :-(for a frown. Excessive and personal criticism of another person in a newsgroup came to be called "flaming," and "flame wars" sometimes erupted, the equivalent of an online shouting match, with reasoned discourse abandoned in favor of name-calling.

The number of Usenet messages exploded in the 1990s, especially after AOL created a method for its millions of subscribers to access Usenet, but the usefulness of Usenet declined in proportion to the number of people using it. The flooding of news groups with advertising messages also drove away many people, who found refuge in e-mail list servers, interactive web sites, and private chat rooms. A major component of the success of Usenet came from the fact that most people did not have access to the Internet; when that access became more common, Usenet no longer offered any serious advantages. The etiquette standards created for Usenet have continued, being applied to BBS chat rooms, web-based chat rooms, and informal e-mail.

GOPHER

As the Internet grew ever larger in the 1980s, various schemes were advanced to make finding information content on the Internet easier. The problem of finding content even existed on individual university campuses, and in the 1980s, various efforts were made to solve the problem on a smaller scale through campuswide information systems (CWIS). Cornell University created their CUinfo, Iowa State created their Cynet, and Princeton created their PNN, all early efforts to organize information.

Programmers at the University of Minnesota released the Gopher system in April 1991 to solve this problem. Gopher consisted of Gopher servers holding documents and Gopher clients to access the documents. The system interface was entirely based on simple ASCII text and used a hierarchy of menus to access documents. The creators of Gopher thought of their creation as a way to create a massive online library. Anyone who wanted to could download the server software, organize their content into menus and submenus, and set up their own Gopher server. Pictures and other multimedia files could be found and downloaded through Gopher, but not displayed within the client. Gopher's virtues included a lean interface and a transmission protocol that did not strain the limited network bandwidth that most users suffered from.

Gopher grew rapidly in popularity, as people on the Internet downloaded the free software and set up their own Gopher servers. Gopher software was rapidly ported to different computer models and operating systems. Even the Clinton-Gore administration in the White House, enthusiastic to promote what they called the "information highway," announced their own Gopher server in 1993. Gopher was the first application on the Internet that was easy to use and did not require learning a series of esoteric commands. Users found themselves enjoying "browsing," going up and down menus to find what gems of text a new Gopher server might offer.

The problem of how to find content reemerged. All these different Gopher servers were not connected in any way, though the Mother Gopher server at the University of Minnesota may or may not have links in its menus to other Gopher servers. Late in 1992, a pair of programmers at the University of Nevada at Reno introduced Veronica. The name came from the Archie comic book series, but in order to make the word into an acronym, they came up with Very Easy Rodent-Oriented Netwide Index to Computerized Archives. Veronica searched the Internet for Gopher files, indexed them, and allowed users to search those indexes through a simple command-line interface. An alternate indexing program from the

University of Utah was called Jughead, again drawing on the Archie comic books.

The number of known Gopher servers grew from 258 in November 1992 to over 2,000 in July 1993 and almost 7,000 in April 1994. In the spring of 1993, the administration of the University of Minnesota, having financially supported the creation of Gopher, decided to recover some of their costs by introducing licensing. The license kept Gopher software free for individual use, but charged a fee for commercial users based on the size of their company. Considerable confusion surrounded this effort, and it sent a chill over the expansion of the protocol. Meanwhile, another protocol, based on hypertext documents, had been introduced to solve the same problem as Gopher.

WORLD WIDE WEB

Tim Berners-Lee (1955–) was born in London to parents who were both mathematicians, and both had worked as programmers on one of the earliest computers, the Mark I at Manchester University. He graduated with honors and a bachelor's degree in physics from Oxford University in 1976. In 1980, he went to work at the Conseil Européen pour la Recherche Nucléaire (CERN), a nuclear research facility on the French-Swiss border, as a software developer.

The physics community at CERN used computers extensively, with data and documents scattered across a variety of different computer models, often created by different manufacturers. Communication between the different computer systems was difficult. A lifelong ambition to make computers easier to use encouraged Berners-Lee to create a system to allow easy access to information. He built his system on two existing technologies: computer networking and hypertext. Hypertext was developed in the 1960s by the development team at Stanford Research Institute led by the computer scientist Doug Engelbart (1925–) and others, based on the idea that documents should have hyperlinks in them connecting to other relevant documents, allowing a user to navigate nonsequentially through content. The actual word "hypertext" was coined by Ted Nelson (1937–) in the mid-1960s.

Using a new NeXT personal computer, with its powerful state-of-the-art programming tools, Berners-Lee created a system that delivered hypertext over a computer network using the hypertext transfer protocol (HTTP). He simplified the technology of hypertext to create a display language that he called hypertext markup language (HTML). The final innovation was to create a method of uniquely identifying any particular document in the world. He used the term "universal resource identifier" (URI), which became "universal resource location" (URL). In March 1991, Berners-Lee gave copies of his new WorldWideWeb programs, a web server and text-based web browser, to colleagues at CERN. By that time, Internet connections at universities around the world were common, and the World Wide Web (WWW) caught on quickly as other people readily converted the necessary programs to different computer systems.

The WWW proved to be more powerful than Gopher in that hypertext systems are more flexible than hierarchical systems and more closely emulate how people think. Initially, Gopher had an advantage in that its documents were simple to create: they were just plain ASCII text files. Creating web page files required users to learn HTML and manually embed formatting commands into their pages. Berners-Lee also included Gopher as one of the protocols that web browsers could access, by using gopher:// instead of http://, thus effectively incorporating Gopher in the emerging WWW.

A team of staff and students at the National Center for Supercomputing Applications at the University of Illinois at Urbana-Champaign released a graphical web browser called Mosaic in February 1993, making the WWW even easier to use. For a time, character-mode web browsers, like Lynx, were popular, but the increasing availability of bitmapped graphics monitors on personal computers and workstations soon moved most users to the more colorful and user-friendly graphical browsers. The WWW made it easy to transfer text, pictures, and multimedia content from computer to computer. The creation of HTML authoring tools made it easier for users to create web pages without fully understanding HTML syntax or commands. This became more important as HTML underwent rapid evolution, adding new features and turning what had been relatively simple markup code into complex-looking programming code supporting tables, frames, style sheets, and Javascript. Berners-Lee's original vision of the WWW included the ability for consumers to interactively modify the information that they received, though this proved technically difficult and has never been fully implemented.

The World Wide Web became the technology that made the Internet accessible to the masses, becoming so successful that the two terms became interchangeable in the minds of nontechnical users. Even technical users, who knew that the Internet was the infrastructure and the WWW was only a single protocol among many protocols on the Internet, often used the two terms interchangeably. Because of slow network speed, graphics-intensive web pages could take a long time to load in the web browser, leading many to complain that WWW stood for "world wide wait."

A major key to the success of the WWW came from generosity on the part of CERN and Berners-Lee to not claim any financial royalties for the invention, unlike the misguided efforts of the University of Minnesota with their Gopher technology. Berners-Lee moved to the Massachusetts Institute of Technology in 1994, where he became director of the World Wide Web Consortium. This organization, under the guidance of Berners-Lee, continues to coordinate the creation of new technical standards to enable the WWW to grow in ability and power. Several times a year, new programming standards for web pages are proposed and adopted, using the RFC system.

An Internet economy based on the WWW emerged in the mid-1990s, dramatically changing many categories of industries within a matter of only a few years. Members of the original Mosaic team, including Marc Andreessen (1971–), moved to Silicon Valley in California to found Netscape Communications in April 1994 with Jim Clark (1944–). Clark had already made a fortune from founding Silicon Graphics (SGI) in 1982, a high-end maker of UNIX computers and software used in 3-D graphics-intensive processing. Netscape brought out one the first commercial web browsers, rapidly developing the technology by adding new features, and became the dominant web browser.

Bill Gates (1951–) at Microsoft recognized that the web browser had the potential to add features and grow so big as to actually take over the role of operating system. This threatened the foundation of Microsoft's success, and Gates reacted by turning his company around from being focused on just personal computer software to an Internet-centric vision. Before this time, Microsoft had concentrated on creating their own online service to compete with AOL and CompuServe, called Microsoft Network. Microsoft was so tardy in understanding the Internet that their first Internet site, a file repository for customer support, was not created until early 1993. Microsoft happened to own their own domain name, microsoft.com, only because an enterprising employee had registered it during the course of writing a TCP/IP networking program.

As part of Gates's strategy, Microsoft released their own web browser, Internet Explorer (IE), offering it for free. Early Microsoft browsers were not technically on par with Netscape, but after several years, IE became a more solid product and Microsoft made strong efforts to integrate IE into their operating system. Doing so allowed them to leverage their monopoly in personal computer operating systems and force Netscape from the marketplace. Netscape was sold to AOL in 1998, mostly for the value of its high-traffic web portal, Netscape.com, rather than the declining market share of its browser. Microsoft's tactics also led to a famous antitrust lawsuit by the federal government, which dragged on from 1997 to 2004, ending with minor sanctions on Microsoft.

Netscape's initial public offering in August 1995 turned the small company into a concern valued at several billion dollars. This symbolized the emerging "dot-com" boom in technology stocks. Billions of dollars of investment poured into Internet-based startups, based on the belief that the Internet was the new telegraph or railroad and that those companies that established themselves first would be the ones that grew the largest. Many young computer technologists found themselves suddenly worth millions, or even billions, of dollars. In such an exuberant time, with speculation driving up stock prices around the world, some pundits even predicted that traditional rules of business had evolved and no longer applied. One of the best examples of dot-com exuberance came in January 2001 when AOL completed their merger with the venerable Time-Warner media company, a deal based entirely on AOL's high stock valuation, which quickly became a financial disaster after AOL's stock value crashed. Alas, in the end, a company must eventually turn a profit. The dot-com boom ended in late 2000, a bursting of the stock market bubble, which caused an economic contraction and depression within the computer industry, and contributed to an economic recession in the United States.

Amidst the litter of self-destructing dot-coms, fleeing venture capitalists, and the shattered dreams of business plans, some dot-com companies did flourish. Amazon.com established itself as the premier online book store, fundamentally changing how book-buying occurred. eBay.com found a successful niche offering online auctions. PayPal provided a secure mechanism to make large and small payments on the web. DoubleClick succeeded by providing software tools to obtain marketing data on consumers who used the WWW, and by also collecting those data themselves. The end of the dot-com boom also dried up a lot of the money that had flowed into web-based advertising. After the dot-com crash, the WWW and Internet continued to grow, but commerce on the web, dubbed e-commerce, grew at a slower rate dictated by prudent business planning.

WEB SEARCH ENGINES

Just as Gopher became really useful when Veronica and Jughead were created as search programs, the WWW became more useful as web-crawling programs were used to create web search engines. These programs prowled the Internet, trying to divine the purpose of web pages by using the titles of the pages, keywords inside HTML metatags embedded within the web page itself, and the frequency of uncommon words in the page to determine what the page was about. When users used a search engine, such as the early www.webcrawler.com and www.altavista.com, a database of results from relentless webcrawling software, sometimes called spiders, returned a list of suggested web sites, ranked by probable matches to the user's search words. Early search engines became notorious for at times returning the oddest results, but they were better results than having nothing.

The other approach to indexing the web was by hand, using humans to decide what a web page was really about. A pair of Ph.D. candidates in electrical engineering at Stanford University, David Filo (1966–) and Taiwanese-born Jerry Yang (1968–), created a web site called Jerry's Guide to the World Wide Web. This list of links grew into a large farm of web links, divided into categories like a library. In March 1995, Filo and Yang founded Yahoo! and solicited venture capital to fund the growth of their company. Thirteen months later, having risen to forty-nine employees, their initial offering of stock earned them a fortune. Yahoo! continued to grow, relying on a mix of links categorized by hand and automated webcrawling.

The web search engine business became extremely competitive in the late 1990s, and many of the larger search engines latched onto the idea of web portals. Web portals wanted to be the jumping-off point for users, a place that they always returned to (often setting up the portal as the default home page of their web browser) in search of information. Web portals offered a search engine, free web-based e-mail, news of all types, and chat-based communities. By attracting users to their web portals, the web portal companies were able to sell more web-based advertising at higher rates.

Larry Page (1971–) and the Moscow-born Sergey Brin (1973–), another pair of Stanford graduate students, collaborated on a research project called Backrub, which ranked web pages by how many other web pages on the same topic pointed to them, using the ability of the WWW to self-organize. They also developed technologies to use a network of inexpensive personal computers running a variant of UNIX to host their search engine, an example of massively distributed computing. In September 1998, Page and Brin founded Google, Inc. The word "google" is based on the word "googol," which is the number 1 followed by 100 zeros. Google concentrated on being the best search engine in the world, and did not initially distract itself with the other services that web portals offered. In this, Google succeeded, quickly becoming the search engine of choice among web-savvy users because its results were so accurate. By the end of 2000, Google received more than 100 million search requests a day.

In 2001, Google purchased the company Dejanews, which owned a copy of the content posted to Usenet since 1995, some 650 million messages in total. This became one of the many new Google services, Google Groups. The success of Google became apparent as a new meaning to the word rapidly emerged, its use as a transitive verb, as in "she googled the information."

7

Computers Everywhere

CONSUMER ELECTRONICS

Computers and electronics have become pervasive in everyday life. What was once high technology is now mundane. For example, on July 1, 1979, Sony offered its first Walkman for sale in Japan. The Walkman, a cassette tape player with headphones, offered the convenience of carrying around your own personal stereo. Sony eventually sold 330 million Walkmans worldwide, including later generations of the ubiquitous object. The consumer electronics revolution made electronics cheaper and provided a new market for computers. In the 1980s and 1990s, video playback machines, video cameras, and stereos all incorporated embedded microchips into their products. Data for electronic devices were usually stored in analog format, but the introduction of audio compact discs for music in the 1980s and DVDs for video in the 1990s showed the advantages of digital recording and playback. Apple introduced the iPod in October 2001, which is the obvious successor to the Walkman, a handheld computer and micro hard drive to play songs recorded in the MP3 format.

Science fiction fans and computer scientists have often dreamed of the day when they had handheld computers. The handheld calculator, based on a microprocessor, partially realized this dream, but people wanted a computer that did more than mathematics. The Palm Pilot, a personal digital assistant (PDA) released in March 1997 by 3Com, fulfilled the dream. Earlier efforts, such as the Apple Newton, had failed in the marketplace because microprocessors were not yet powerful enough and people demanded effective handwriting recognition. After the success of the Palm Pilot, other companies entered the PDA market.

In the mainframe era, one computer serviced many people, then the personal computer revolution offered the potential to give each person a computer, and with what some have called ubiquitous computing, in the 1990s, one person now had many computers to serve him or her in a variety of roles. A digitally oriented person today may use a desktop PC, numerous server computers through the Internet, a PDA, and an MP3 player. In addition, non-computer-oriented people use watches, stereos, televisions, automobiles, cell phones, pagers, microwaves, and other electronic equipment, all equipped with embedded microprocessors.

DIGITAL DIVIDE

In the mid-1990s, observers concerned about the importance of computer use began to fret about what they called the "digital divide." Another way to phrase the problem was that some people were "information rich" and other people were "information poor." Because of poverty and the lack of opportunity, most of the 6 billion people on the planet did not and still do not have access to computers.

A set of studies in 2000 found that more than half of the computers on the planet were found in the United States, where 51 percent of U.S. families owned a computer and 41 percent of U.S. homes had Internet access. Western Europe was also well connected, with 61 percent of Swedish homes having Internet access, though only 20 percent of Spanish homes had Internet access. Thirty-three percent of Asian homes had Internet access, especially concentrated in Japan, South Korea, and other up-and-coming economic powerhouses. Many nations, especially in Africa, had only minuscule numbers of computers and limited Internet access. These statistics illustrated the global digital divide, where citizens of affluent countries had access to computers and Internet access, and where poorer countries were not able to provide the education and infrastructure to compete in an increasingly globalized economy.

The same set of studies found that in the United States, 46 percent of white Americans had Internet access at home, while only 23 percent of African American and Hispanic homes had access. Eighty-six percent of households that earned more than $75,000 per year had Internet access, while only 12 percent of households that earned less than $15,000 per year had access. Sixty-four percent of households with a college graduate had Internet access, and only 11 percent of households had access if no one in those homes had ever graduated from high school. Urban and suburban homes were more likely to have Internet access than rural homes or homes in impoverished inner cities. This showed that the digital divide also existed within nations, a digital divide exacerbated along racial, income, and educational divides.

Observers expressed concern both with access to computers and the Internet, and with knowledge of how to use computers and the Internet. Some have argued that computer literacy is almost as important as traditional literacy in order for a person to fully participate as a citizen in a modern democracy and to even find a good job. Various efforts have been made to bridge the digital divide, concentrating on educational efforts and government programs to provide more universal access to the Internet, similar to earlier programs to provide universal access to basic telephone service.

Most national governments have been concerned about regulating Internet content, even if only to prevent the spread of child pornography. Some repressive governments, recognizing the importance of computers and Internet access to the success of their economies, encourage computer and Internet use, but strictly regulate content. For instance, Internet access is common in Chinese cities, but all Internet providers pass their networks through official Chinese censorship filters. A person walking into an Internet café in China cannot access any Western news sites, sites set up by Chinese dissidents or human rights organizations, or Western pornography. Visitors to China can use Internet connections that bypass the Chinese censorship filters in hotels that cater to foreigners. One of the more interesting social implications of networked computers will be to see if computers and the Internet will erode the power of repressive governments to restrict information flow or if those governments will succeed in their efforts.

THE OPEN SOURCE MOVEMENT

The open source paradigm began to revolutionize the computer industry in the 1990s. The majority of the world's web sites, including many of the biggest web sites in the world, such as Google, Amazon.com, and Yahoo!, run on open source software. Open source software is software that is published with the actual source code included, instead of in a format that allows just execution of the program. For example, when you purchase Microsoft Office, you can install and run the program, but you are not given the proprietary source code that allows you to alter the program and change it according to your own needs. Most open source programs are distributed for free, with programmers contributing their time because they feel passionate about the programs.

In the 1970s, software became so important that new companies sprang up to solely produce software. Software became big business, spawning the economic giant Microsoft (founded in 1975) and a host of other companies. Software proved to be wildly more profitable than hardware, with profit margins of up to 80 percent. Because software was prone to be buggy, legal contracts called end-user license agreements (EULAs) became common to prevent customers from suing because the software did not perform as advertised. Because the software industry does not actually sell their products, but actually just licenses their use, and because of the EULAs, the software industry is unique in being able to sell defective products with minimal legal consequences.

Operating system software during the 1970s also become more sophisticated. While most operating systems remained proprietary software, running only on the hardware also sold by the manufacturer, AT&T Bell Labs developed an operating system called UNIX in the early 1970s. UNIX became popular on university campuses and became unique in that the same operating system soon ran on many different hardware platforms, regardless of the manufacturer. UNIX also become the operating system of choice for the engineering and graphics workstation market that flourished in the 1980s.

Richard Stallman (1953–), a programmer at the Massachusetts Institute of Technology (MIT) Artificial Intelligence Lab, enjoyed sharing software that he wrote with other users and using the software that they shared with him. In 1984, inspired by his ideal that software should be free, Stallman refused to join the burgeoning software industry, but quit MIT to create the GNU project. GNU is a recursive pun, meaning "GNU Not UNIX." Stallman's crusading project aimed to recreate the UNIX operating system and the common tools on UNIX from scratch so that the GNU programs would be free from copyright and anyone could use them. Stallman and some law professors created the GNU General Public License (GPL), which Stallman characterized as a copyleft scheme rather than a copyright scheme. All GPL'd software must be released as free software with source code included. The software is not in the public domain, but remains copyrighted so that it may not be readily used in commercial software. The GPL is characterized as "viral" in that any new software that includes any GPL'd software within it must then be released as GPL software. The GPL is not the only open source license, but it is widely used.

In 1985, Stallman and like-minded programmers created the Free Software Foundation (FSF), funded through sales of GNU programs and donations. By selling software, the FSF seemed to violate its own philosophy of developing free software, but it is really just selling the disks and manuals, not the software, which, since it is licenced under the GPL, can be legally passed on to anyone else. The FSF created many of the common tools on UNIX, such as a compiler and text editors, but it remained for Linus Torvalds (1969–), a graduate student in Finland, to create the Linux operating system in 1991, a clone of UNIX that used the GNU tools. Other open source UNIXes also appeared, such as FreeBSD and NetBSD. Open source projects often concentrated on the UNIX operating system, but many projects were quickly ported to other operating systems, including the Microsoft Windows family.

While the founding of the FSF was a reaction to the commercialization of software, few people shared Stallman's idealistic vision, and copyright and patent rights were not abandoned by the open source movement. Linux vendors such as Red Hat and MandrakeSoft figured out in the 1990s how to make money from open source software by selling convenience in obtaining updated software, service contracts, and consulting. Other vendors, such as MySQL, which marketed an open source database, provided a choice to customers. If a customer wanted to use the MySQL database software as part of another software product, then the customer must either pay a licensing fee or make their product free and open source.

Open source grew quickly during the 1990s and early 2000s, becoming an important market force. Open source software was created in three ways:

- Amateur projects
- Important projects sustained by competent professionals donating their time
- Projects supported by companies who found that open source gives them certain competitive advantages

Many times, if the product were important enough, volunteers and company employees worked on the same open source projects. Examples of this included Linux, the GNU tools, and the Apache web server.

The open source movement represented more than idealism and programmers doing what they love. Once programmers see that an idea is possible, they are halfway to replicating that idea. This means that proprietary software is forced by competition from similar open source products to continually improve, selling products with better features because the older features are becoming part of the common reservoir of software ideas.

Open source projects became possible because of the Internet. Spread around the world, open source programmers communicate via networks, bound only by their common interests in a project and striving for excellence. Projects are usually driven by a small group of core programmers, and decisions are usually made by consensus. Programmers rise to leadership positions because of their recognized skills, a mark of high-technology personal status and a form of pure meritocracy. Some programmers are much more productive than other programmers. These uber-programmers are often twenty times more productive than the average programmer. It is these programmers, the cream of an elite, who are the backbone of both important proprietary and open source projects.

As programmers became more experienced with solving a particular type of problem, they create ever more sophisticated algorithms, or recipes for getting programs to do things. As time has gone by, the community of software developers has come to understand certain types of software so well that writing that type of software is a straightforward process. Text editors, operating systems, word processors, office productivity suites, e-mail clients, media viewers, web browsers, web servers—all have solid open source products. This is the commoditization of software.

Ironically enough, considering that open source is often not backed by the financial resources of a large corporation, open source software has a justly deserved reputation of being less buggy and more reliable than proprietary commercial software. This strange paradox occurs because software engineering researchers have found that one of the best methods to remove bugs from a software product is code review, where multiple programmers look over the code, visually inspecting the code for flaws. The open source process, by its very nature, where anyone can look at the code, has more eyes on the code than comparable proprietary software teams.

Programmers involved in open source projects are driven by a passion for excellence. Open source programmers put technical merit before market success. Unlike commercial companies, who will ship products to meet a deadline, open source projects only declare a product ready when it is really ready. Early releases of open source products happen, but are publically labeled as alpha versions, not complete products. High-quality software is intrinsic to the open source movement, though the movement is littered with products that have been abandoned because programmers lost interest. This is not a problem since these products have become obsolete or did not attract sufficient users to maintain programmer excitement, a form of natural selection. When a product type is ripe to be commoditized, an open source version will appear.

Open source products present a seeming paradox for computer security. Some pundits argue that because the source code is visible to everyone, then malicious hackers can work their way through the code line by line, finding a bug or an oversight that can lead to a way to compromise the open source product. This fear is certainly true, though experience has shown that flaws in closed source products are more common because not as many programmer eyes look at the source code for the product. After a period of seasoning, open source products are stable and secure.

The open source Apache web server is now used by the majority of the world's web servers. Figures from the computer industry research firm Gartner show that Linux is the fastest-growing operating system in the world. Gartner further predicts that by 2007, Linux may account for 15 percent of the global market. IBM and Hewlett-Packard already make billions of dollars off of open source–related products. Apple Computer's newest operating system for their Macintosh computers, Mac OS X, based on the NeXT technology, uses an open source flavor of UNIX as its foundation.

Linux and the other open source operating systems mostly run on Intel-based microprocessors because they are cheap to purchase and support. In 2003, more Linux servers were shipped than servers with proprietary UNIX operating systems, part of a process of open source operating systems coming to dominate the UNIX market, forcing the proprietary UNIX products out of the market, such as Solaris from Sun Microsystems, AIX from IBM, and HP-UX from Hewlett-Packard. Linux and other open source products were making their greatest strides in the enterprise arena in the server rooms of large and small businesses. Though open source desktop products were growing stronger, they had not reached prominence in the minds of common computer users.

Agencies of local and national governments in Germany, France, China, Britain, Israel, Japan, Brazil, South Africa, Russia, and South Korea have announced their intention to use open source software or are seriously investigating converting to an open source operating system and applications. This development has so concerned Microsoft that they have created a special program for foreign governments to be able to look at the Windows source code. Microsoft misses the point here. Government information system professionals do not just want to look at the code; they also want to make the code better or change the code to suit their particular needs.

Even though Microsoft Windows and Microsoft Office run on about 90 percent of the world's desktop computers, Microsoft has identified Linux and open source projects as a major risk to their future business. Microsoft was founded on the premise that people will pay for personal computer software, and early personal computer hobbyists in the mid-1970s resented the small company for charging for their products when everyone else gave their products away as part of the old hacker culture. Microsoft proved to have found the correct paradigm for the past three decades, but open source promises a possible new paradigm. In the future, open source software will probably become ever more important. The enthusiasm of programmers is the powerful currency propelling open source, and Linux and other open source projects will continue to grow because there is no company to fail and no profit margin to maintain.

HACKING AND INFORMATION SECURITY

We all have a mental image of hackers derived from the news media, movies, cyberpunk novels, or friends that we know. The term "hacker" emerged in the 1960s, when writing computer code was a difficult, esoteric art and computer experts were called hackers as a badge of honor. Some of these hackers used their skills to enter systems without permission. Another breed of technical wizards emerged in the 1960s with phone phreaking. Phreakers learned how to manipulate the AT&T phone system to avoid bills, make prank calls, and make free long-distance calls. Cap'n Crunch (a.k.a. John Draper), the most famous of the phreakers, earned his name when he discovered that the whistle that came in a box of Cap'n Crunch cereal emitted the right frequency to subvert the AT&T phone system. At that time and even today, telephones were based on acoustic technology and responded to different tones to know what numbers to dial, if enough change had been deposited into a payphone, or how to activate the repair mode that telephone repair personnel used to troubleshoot problems. Cap'n Crunch reputedly taught Steve Wozniak how to be a phreaker and the Woz got his friend, Steve Jobs, involved in their first business—building electronic "blue boxes" to sell to other phreakers in order to cheat the phone company—before the Woz and Jobs later turned honest and founded Apple Computer.

As time went by and some hackers started to cause problems, the term "hacker" changed from a moniker of respect to a label defining an antisocial person bent on making trouble. Some hackers, wanting to regain the title, suggested that we instead use the term "cracker" for the bad guys, but that never caught on. Some alternate terms now used are "white hat" hacker and "black hat" hacker, recalling the days of cowboy movie serials where the audience never doubted who was good and who was bad. Another common term is "script-kiddies," novice hackers who only know how to run hacking tools and do not understand the underlying technology.

Many hackers are motivated by the desire to seek knowledge and to be respected for their technical prowess, even if their real identity is hidden behind a pseudonymous handle or they are part of a larger hacker organization. Serious hackers enjoy conquering difficult technical problems, and are often thrill-seekers or are seeking revenge. Laid-off employees are a major source of hacking. The best hackers often seek financial gain, either through industrial espionage or outright theft. Multiple episodes have been reported where hackers steal credit card information from a bank or online retailer and blackmail the targeted company for the return of the information. Companies often pay the blackmailer because they want to avoid a public relations disaster and loss of confidence by their customers. Another motivation for hackers is hactivism, where they hack for a social or economic cause, often defacing web sites of companies or organizations that they disagree with.

So how do hackers do it? Quite frankly, social engineering is the most effective method. A hacker calls up someone in a company, poses as a technician from the internal computer support department, and asks for a password. Most of the time, the hacker will get the password through this simple ruse. One of the most famous social engineering hacks occurred in 1978 when a computer consultant named Stanley Mark Rifkin performed a temporary assignment at the Security Pacific National Bank in Los Angeles. He learned how the system of wire transfers worked and realized that security centered around a daily code that was given over the phone by bank executives to authorize a wire transfer. He visited the wire transfer room, saw the code written on a piece of paper, then went outside the bank to a payphone. When he called the wire transfer room, he pretended to be a bank executive and asked for $10.2 million to be transferred to a bank account in Switzerland. When asked for the code, he gave them the number. Rifkin then flew to Switzerland and converted the cash in his account to diamonds before returning to the United States to enter the *Guinness Book of World Records* with the record for the "biggest computer fraud," though this record is no longer awarded.

Hackers might also use a program called a port scanner to knock on all the doors on a network or computer system to see if any points of entry are willing to communicate. After gaining communication with a system, the hacker must still use his or her bag of tricks to try to gain access to the system. Besides having their own custom software, hackers often search the many databases available on the Internet for an exploit that opens the vulnerable hole wide enough to gain control of the target system. The best hackers come and go with no one ever being the wiser.

So how bad is the information security problem today? There is a lot of hacking going on, with lots of worms and viruses roaming around in the wild, ready to take advantage of any unprepared machine. Most people who have a persistent Internet connection (not a dial-up modem) experience dozens or hundreds of exploratory probes per week. Most successful hacks go unreported because either the person or organization fears bad publicity or they have no idea that they were hacked. The Internet, by connecting computers around the world together into a massive cybernetic organism, makes it easy for hackers to hit computer systems or networks from another continent.

In 1997, the NSA authorized a team of their hackers to penetrate American military and civilian computers in an exercise called Eligible Receiver. It took several days before the unsuspecting military believed that they were really under attack. While the particulars have not been released, apparently the hackers experienced considerable success, leaving messages on systems they had compromised and taking over command center computers, systems on power grids, and the 911 emergency call systems in nine American cities.

In 2000, as part of the security effort for the 2002 Winter Olympics in Salt Lake City, Utah, the U.S. Department of Energy and Utah Olympic Public Safety Command ran an exercise called Black Ice. Hundreds of participating officials tried to cope with a major ice storm that damaged power lines and resorted to a policy of rolling blackouts. Officials were surprised at how much intermittent power degraded their infrastructure and affected their ability to communicate in an emergency. While the scenario did not deal with cyberterrorism, the effects demonstrated how interconnected the various systems of the public infrastructure are and how vulnerable they are to malicious attack. A real example occurred that same year when, after at least forty-six attempts to break into a computerized sewage system in Maroochy Shire, Australia, a disgruntled former employee succeeded and released 1 million liters of raw sewage into local parks and streams.

The most effective attacks come not from the brute force of crashing systems, but from subtle changes in data on computer systems that lead people to make bad decisions. Possible attacks on digital control systems for gas pipelines, electrical power grids, hydroelectric dam controls, sewage treatment plants, water distribution systems, oil refineries, or chemical manufacturing facilities can also cause major utility disruptions or environmental damage.

The key to whether a computer system is vulnerable is whether the hacker can communicate with the system. A hacker must interfere with a wireless signal (found in links with satellites, or by using microwave, infrared, or radio waves), or have access to the actual communications wire that is attached to a computer system. The American military and different government agencies have separate computer networks from the open Internet, though even these private networks are often created by leasing network bandwidth from large national and international providers like AT&T, MCI, Global Crossing, or Sprint. The only way for a computer system to be completely safe from a hacker is to not have a network and keep the computer in a locked room.

Most people know someone who has caught a computer virus or worm. These terms are often mixed up and used improperly. Just as in biology, a virus needs a host to live in and reproduce, whereas a worm can reproduce on its own and travel about on its own. Perhaps the most famous of all worms, the Morris Worm, happened in 1988. Robert Morris Jr., a graduate student at Cornell University who wrote the worm, did not intend for it to damage anything, but just wanted to see how far his worm would spread. A bug in the program allowed the worm to keep reinfecting systems and eventually caused systems to slow down and sometimes crash. The Internet was much smaller at that time, and Morris infected an estimated 10 percent of all machines on the Internet. The federal government convicted Morris of his crime and sentenced him to three years of probation and a fine. Since the mid-1990s, the public has become increasingly aware of the impact of computer viruses and worms with exotic names such as Slammer, Nimba, Code Red, and Melissa. Most of these viruses and worms are relatively unsophisticated efforts, though they have caused the loss of millions of dollars in lost computer time and in the effort needed to clean up afterwards.

For information security professionals, the "white hats," the best defense comes from knowing your enemy. Some people have proposed counterattacking any attacking machine, getting the hacker before he or she gets you. This is not a wise idea because any hackers worth their weight in salt launch their attacks from computer systems that they have already compromised, and the true owner of the system is unaware of what is going on. Security professionals, sometimes called "ethical hackers," always get permission from the proper authorities before doing a security analysis of a computer system or network. This permission is referred to as a "get out of jail free card," because would-be security analysts have gone to prison while claiming in their own defense that they just wanted to show an organization what its security problems were.

Most people think about security as an afterthought. Good security means an absence of problems, an intangible that makes it hard to justify spending extra time or money on more diligence. Many people, even those that should know better, follow the "security through obscurity" philosophy. If they just keep quiet about their poor security, then perhaps no hackers will notice them. Hackers will eventually find any target of easy opportunity that is connected to the Internet, so competent organizations have firewalls that examine network traffic and restrict access to only authorized data or programs. They also run intrusion detection systems (IDS), guardian programs that monitor networks and systems for suspicious behavior.

Encryption is often used to help enhance information security. Just as the bombes and Colossi in Britain during World War II used early computers in decoding efforts, computers have remained important in later advances in encryption and decryption. Electromechanical devices like the Enigma and electronic computers made possible using more complex encryption and decryption algorithms than if we relied only on calculations by hand.

Before the 1970s, all code algorithms were symmetric, in that they used the same algorithm and key to both encode and decode a message. In 1975, cryptographers Whitfield Diffie (1944–) and Martin Hellman (1945–) developed the mathematics for public-key encryption, an asymmetric algorithm. Two years later, Ron Rivest (1947–); Adi Shamir (1952–), an Israeli; and Leonard Adleman (1945–) developed and patented the RSA algorithm (named after their initials), an effective implementation of public-key encryption that became a standard in the field. In its simplest form, public-key encryption allows for the creation of a pair of keys that can then be used to encode or decode a message. Only the opposite key can decode what the other key has encoded. Possessing only one key does not allow someone to break the encryption.

In the last decade, computer security has become ever more important. Hackers, viruses, worms, and Trojan horses are a major concern for computer users everywhere. Trojan horses are malicious programs disguised as harmless programs. In the past, when a customer asked if a product was secure from hackers, the software company often said something like, "Of course it's safe. Our encryption algorithms are secure because we don't show them to anybody. We keep everything secret." Security professionals now recognize that security through obscurity does not work. The best encryption algorithms in the world are public knowledge, where anyone who wants to can analyze them for weaknesses. It is this process of review that demonstrates how strong the algorithms really are. Companies who use secrecy as a substitute for security are engaged in lazy software engineering.

Movies that portray hackers in a technically accurate way include *WarGames* (1983), *Antitrust* (2001), and some aspects of the *Matrix* movies. Truly awful hacker movies include *Hackers* (1995), *Swordfish* (2001), and most movies that portray hacking as a way to advance the plot. *The Net* (1995) and *Sneakers* (1992) come down in the middle, telling engaging stories but using an implausible decryption technology that behaves like magic.

CYBERWAR

On June 3, 1980, a microchip costing only 46 cents failed, generating an alert that a Soviet submarine had just launched two nuclear-tipped missiles at the United States. The Americans immediately reacted by automatically raising the alert level on their nuclear forces, sending 116 bomber crews to their aircraft with orders to start their engines. American submarines carrying their own nuclear-tipped missiles were alerted. Within only a few minutes, the American commanders realized that the alert was false, but the episode accentuated how dependent they had become on computer technology to give them the edge in the split-second timing demanded by possible nuclear war, and a sharp lesson that computers could fail.

Cyberwar, also called "information warfare" or "netwar," is vaguely defined, but all definitions include using computers as tools of warfare, both for defense and offense. Cyberwar grew out of two twentieth-century developments in warfare. As troops grew to rely on radio, radar, sonar, and other electronic sensors, enemy forces developed electronic countermeasures (ECM) to confuse those electronic sensors and even render them useless. This might be as simple as jamming a radio frequency with noise so that your adversary cannot use that frequency for radio transmissions. For that reason, modern soldiers use frequency-hopping radios that move so quickly from frequency to frequency that the enemy cannot keep up and adjust their jamming equipment quickly enough. The effort to counter ECM has its own acronym, electronic counter-countermeasures (ECCM), and the war within the electromagnetic spectrum is a never-ending cycle of creating new attacks and new defenses.

The second development included ever more sophisticated communications, command, control, and intelligence (C3I) military infrastructures. Modern warfare, as practiced by the American military, relies extensively on computers and electronics. Satellites spy on the enemy and transmit encrypted communications, computerized databases streamline logistics, and the American military is now deploying electronics where each soldier is literally turned into a node (system) on a network. The soldier will be able to feed video back to his commander, receive orders, and view a video feed from small overhead unmanned flyers. The ultimate cyberwar weapon is a nuclear device modified to maximize its electromagnetic pulse (EMP), which can literally destroy running electronics and electrical systems.

The requirement for the American military to always be able to precisely locate their personnel and equipment anywhere on the Earth's surface led to the American Global Positioning System (GPS). This system of twenty-four satellites, launched between 1989 and 1994, now allows anyone with a GPS receiver to locate him or herself on the surface of the Earth with accuracy of closer than 10 meters. American military users have better resolution. GPS equipment attached to missiles and bombs has made precision weapons easy and effective, seen most prominently in the recent wars in Afghanistan and Iraq. Ever since the 1920s, aerial bombardment enthusiasts have expected air power to become the decisive weapon on the battlefield, and GPS-enabled precision weapons may have actually fulfilled that promise. Anyone can use the GPS signals, and GPS receivers have revolutionized the practice of scientific fieldwork by allowing precise measurements of continental drift and the location of geological formations, archaeological sites, and animal populations. GPS receivers are used by hikers, to locate stolen cars, and to track commercial shipments. Potential enemies can also use GPS receivers, and the American military is rumored to have a feature that, if necessary, will change the GPS signals so that only American military GPS receivers would continue to work and all civilian receivers would fail.

Many American federal agencies have overlapping responsibilities on the issue of cyberwar, including the Pentagon, the National Security Agency (NSA), the Federal Bureau of Investigation, and the Central Intelligence Agency. The NSA, nicknamed the "puzzle palace," is perhaps the most misunderstood of the agencies. The NSA, founded in 1952 by presidential order (the only federal agency founded without congressional action), is the codemakers and codebreakers for the federal government. The goal of the NSA is to conduct electronic surveillance around the world, break the codes if encrypted messages are found, and create encryption schemes for use by the American military and government that cannot be broken. The NSA and similar organizations in Britain and Australia set up the Echelon system to intercept e-mail, fax, telex, and telephone communications around the world. While the NSA denies that Echelon exists, an inquiry by the European Parliament asserted that it does exist. The NSA is also rumored to have more supercomputers than any other organization in the world, to better make and break codes.

A major concern of cyberwar theorists is not just the potential effects on the battlefield, but also the potential for a "Pearl Harbor" cyberattack on civilian targets by cyberterrorists or other enemies. Less dramatic scenarios are also easy to imagine. What if you go to your ATM and can't withdraw any money because a bank computer system thinks that your balance is zero; you can't fly because the air traffic control system has crashed; you can't leave your sorrows behind in a game of Everquest because the Internet is clogged with rampaging worms, viruses, and denial of service attacks; and you have no electricity because the computer systems controlling the electrical grid have shut down? Or imagine soldiers under attack by surface-to-surface missiles unable to fight back because their missile defense computers have suddenly rebooted because of a virus, or a naval ship dead in the water because its integrated control is confused by buggy software.

ARTIFICIAL INTELLIGENCE AND ROBOTICS

In the early 1980s, expert systems showed promise. An expert system is created as a series of rules, with an inference engine, covering a narrow field of expertise. The first successful commercial expert system configured orders for Digital Equipment Corporation (DEC) computer systems. The promise of expert systems created a commercial expansion of the field, only to find disappointment by the end of the 1980s as the expense and limitations of expert systems became apparent.

Artificial intelligence (AI) researchers developed a close relationship with cognitive science, the study of how humans think. One result of this relationship was neural networks, implemented in either hardware or software, which simulate the way that the human brain uses neurons to form patterns and to change the relationships within those patterns. The concept of fuzzy logic also grew out of this relationship, allowing software to make decisions when input data remained uncertain and incomplete.

Robotics was related to AI in that many of the difficult problems in each field are similar. Industrial robots that perform limited tasks became common in advanced factories the 1980s, while autonomous robots that can correctly perceive the natural world through vision or other sensory means, and react to that sensory data, remain the dream that drove robotics research. Initial efforts in robotics quickly revealed that what humans or animals did naturally, with no effort, was very difficult. A robot called Shakey in 1969 at Stanford University required 5 minutes to move a foot even though the room contained few obstacles. The computer processing power required by the programming of Shakey could not be carried aboard the 5-foot-tall robot, so an external computer was attached to the robot by a cable. Many researchers thought that the problems of programming autonomous robots would be solved with ever more powerful computers, but that proved to not be true. Significant work at MIT's AI Lab in the 1990s developed a different bottom-up approach to robotics, creating machines that move like insects using simple algorithms.

Vision is a form of pattern recognition, and AI research eventually led to pattern recognition of handwritten characters, which became more common by the late 1990s. Effectively recognizing handwriting patterns also became possible in the 1990s. The related problem of voice recognition has resulted in simple commercial applications.

Some of the more glamourous successes of robotics came from the American Mars missions of the National Aeronautics and Space Administration (NASA). The Mars Pathfinder mission in 1997 landed the robotic rover called *Sojourner*, which crawled for short distances under remote control from Earth. In 2004, two further rovers were successfully landed to search for geological evidence of water and past life, and they traveled up to 100 meters a day in a semiautonomous mode.

In 1997, an IBM supercomputer called Big Blue defeated Garry Kasparov (1963–), the world chess champion, in a chess tournament. Kasparov later complained that Big Blue had been programmed to specifically defeat him, though all grand masters train to defeat specific opponents. Many commercial computer games and video games have rudimentary AI algorithms to provide an artificial opponent for the human player, with decidedly mixed success at being a challenge. Many canny players find the conceptual blind spots in the game's AI and ruthlessly exploit them to their advantage.

FINAL THOUGHTS

Science fiction is the literature of the scientific age. From research on atomic weapons to understanding DNA, the evolving computer has been both the goal and critical component of our scientific age. As such, it has been a critical component of the literature that seeks to describe and even promote scientific advances. Current research and development with computing promise even more tantalizing science fiction futures, both obvious and scarcely imagined.

Born in adventure pulp magazines during the first half of the twentieth century, the science fiction genre began to mature in the 1940s and 1950s. While many of the early stories suffered from simplistic characters and juvenile plots, speculations on the impact of science and technology, and extrapolations of possible future directions of science and technology, have proved to be insightful and useful. In the early 1980s, computers inspired the cyberpunk movement in the genre. Though the rise of personal computers was almost completely absent from science fiction prior to the 1970s, cyberpunk fiction made up for that, describing worlds where computers and the flow of data completely transformed human relationships and the common reality that everyone experienced. Often this cyberpunk vision turned depressing and pessimistic, though the 1990s has seen cyberpunk themes absorbed into the mainstream of science fiction. The possibilities of nanotechnology also excited the imaginations of science fiction authors in the 1990s.

Research on manufacturing objects on the scale of 100 nanometers or less is called "nanoscience" or "nanotechnology." The 1986 book by the molecular nanotechnologist K. Eric Drexler (1955–), *Engines of Creation: The Coming Era of Nanotechnology*, popularized the term "nanotechnology" and introduced his vision of a world dominated by nanoengineering. Drexler first published a paper on molecular manufacturing systems in 1981, drawing on the ideas of the Nobel laureate physicist Richard P. Feynman (1918–1988). Drexler later earned the first doctorate in molecular nanotechnology in 1991 from the Massachusetts Institute of Technology (MIT), where his dissertation supervisor was the artificial intelligence pioneer, Marvin Minsky (1927–).

By building microscopic objects atom by atom, nanoengineers hope for revolutionary advances in the material sciences, leading to advances in computers. Nanotechnology will only fulfill its promise with sophisticated new programming techniques, probably drawn from a better understanding of the genetic programming found in natural microorganisms. The ultimate dream is to create nanobots, microscopic robots with the ability to manipulate matter on a molecular level. The next step is replicating assemblers, generic nanobots that can create more copies of themselves. Medicine might benefit from advanced sensors for disease detection, new therapies, and new implants, and by using nanobots to operate on internal organs or repair the interior of human cells. In 1989, the IBM physicist Don Eigler used a scanning tunneling microscope to arrange thirty-five xenon atoms to spell the microscopic word "IBM." More practical examples of nanotechnology emerged in the 1990s, with molecular-engineered nanoproducts eventually making up a worldwide market of over $40 billion. This includes older products such as microchips and newer products such as carbon nanotubes.

Vigorous advocates of the potential for nanotechnology often paint scenarios of future possibilities that are derided by critics as mere science fiction, and indeed, science fiction authors in the 1990s and afterwards often used nanotechnology extensively in their stories, often as a kind of

Scanning tunnelling microscope photograph of the word IBM spelled in xenon atoms. IBM Corporate Archives.

magic. Arguments similar to those opposing genetically modified organisms are also offered against nanotechnology. Critics of nanotechnology fear that nanobots that can self-replicate might run out of control, like a bacteria run amok. What mechanisms in nature might stop nanobots? Nanotubes, microscopic tubes made of carbon atoms, are a major advance in nanotechnology with many applications. Researchers at DuPont injected nanotubes into the lungs of rats, and 15 percent of the rats died within a day from blockage of their airways. Studies such as the DuPont study lend credence to the arguments of critics.

In the 1990s, the National Science Foundation (NSF) coordinated nanotechnology research and funding under a National Nanotechnology Initiative. The National Institutes of Health (NIH) also funded nanoscience and nanotechnology research project grants in biology and medicine through the NIH Bioengineering Consortium. Research in nanotechnology gained a boast, becoming a big science project, when the U.S. Congress allocated $3.7 billion in 2003, to be spent over a 4-year period, on nanotechnology research. This funding aimed to establish a network of university-based technological research centers, and included an effort to also fund research into the ethical issues that might arise from nanotechnology. Observers expect many of the ethical issues to be similar to issues already familiar from the field of bioethics.

The preceding descriptions of possible technologies and scientific advances barely scratch the surface of the possible futures created by advances in computing. While narratives of progress are often regarded with suspicion by historians and other scholars, few will argue that the story of the rise of information machines is a story of technology growing ever more sophisticated and ubiquitous. Even more than the printing press, the computer has opened up global communications to multitudes. By improving the means by which we communicate, entertain ourselves, travel, calculate, and do a thousand other things, the computer will be an essential tool in reaching for the stars and our quest for all that might follow.

The computer has also accelerated the pace of technological change, so much so that some pundits predict a singularity in the near future when computers completely transform humanity's ability to manipulate reality. Science fiction may imagine these things. With computing, reality is not far behind . . . and sometimes, it is even ahead of our imagination. Certainly, a reader fifty years from now will look back on the computers and software available at the turn of the millennium and be astonished at how primitive it all is.

Glossary

Algorithm. A set of instructions to perform a task on a computer, like a recipe.

Analog. The Antikythera device described in Chapter 1 is known as an analog device, while the modern electronic computer is a digital device. A digital device has built-in stops. For example, the modern computer has discrete clock cycles, the screen image is composed of many discrete "pixels" that are turned on or off, and the sound is composed of discrete sampling of the original sound. Digital modeling can do a fair representation of the rounded shapes and sounds of reality; however, human ears and eyes can still distinguish between analog and digital representations. Analog devices have had a distinct advantage over digital devices over time and predate digital devices because of those advantages. For one thing, analog devices can fill in the blanks, the holes between discrete samples. For another, analog models of reality can be created without creating the often complicated basic mathematical equations of reality to be modeled. Analog machines are specialized for particular purposes, not general machines made for many purposes. For this reason, analog computational devices remain with us—a simple example is the analog gas meter measuring the gas used by a home.

Architecture. The design of a computer system, both its hardware and software. The term can also refer to the design of a computer program.

Assembly language. A level of programming above machine code, where instructions are identified by easier to remember mnemonics, such as JMP for the Jump instruction.

Basic input/output system (BIOS). The programming code, usually found on a ROM chip, that contains the basic input and output routines for peripheral devices, such as the floppy drives, hard drives, keyboard, or monitor.

Bit. A single binary value: on or off, 0 or 1.

Bitmapped graphics. A method of mapping the contents of RAM memory onto the pixels of a cathode ray tube (CRT) screen.

Bus. A set of wires that creates a data path to allow different components in a computer to communicate with each other.

Byte. 8 bits of memory, large enough to hold a single character in the English language. 1,024 bytes is a kilobyte, 1,024 kilobytes is a megabyte, and 1,024 megabytes is a gigabyte.

Cathode ray tube (CRT). A display device used for early memory storage and terminal screens.

Central processing unit (CPU). The main processing component of a computer. Microprocessors are a complete CPU.

Digital. *See* the explanation under Analog.

Electromechanical. A system or device that combined the electrical flow of electrons with the physical movement of mechanical parts.

Hardware. The physical components of a computer system, as opposed to software.

Integrated circuit. An electronic device, based on silicon, consisting of numerous electronic components, such as transistors, diodes, resistors, capacitors, and inductors.

Local area network (LAN). A network allowing computers to communicate within a building. *Also see* Wide area network (WAN).

Logarithms. An arithmetic series of numbers that correspond to another geometric series of numbers. The correspondence makes calculating multiples of geometric series easier by reducing the calculation to an addition of the arithmetic series and a lookup of the corresponding geometric value. For example, take the geometric series of numbers where each number is two times the previous number: 1, 2, 4, 8, 16, 32, and so on. Each number in the series corresponds in order to its logarithm in an arithmetic series: 0, 1, 2, 3, 4, 5, and so on. If we want to calculate multiplying two numbers from the geometric series, 4 times 8 for example, we add the corresponding numbers in the arithmetic series: 2 plus 3, which gives us 5. The geometric number corresponding to 5 is 32, the answer to 4 times 8.

Machine code. The binary code that makes up all computer programs. Higher level languages, such as assembly code, Fortran, COBOL, Pascal, C, or Java, must be compiled or interpreted into machine code before the program can actually run on the computer.

Mainframe. A large computer whose size and features have changed during the last six decades, but usually were the biggest computers available outside of the specialist category of supercomputers.

Microchip. An integrated circuit.

Microcomputer. A small computer based on a microprocessor, intended for the use of a single user.

Modem (modulator/demodulator). A device hooked between a telephone and a computer to convert to and from an analog signal that the telephone system will understand to the digital signals that a computer understands.

Multiprogramming or multitasking. The ability to run more than one program at a time on a single central processing unit (CPU) by dividing time on the CPU between the different programs, also called tasks.

Network. A set of computers connected together to exchange data, usually either a LAN or a WAN.

Network bandwidth. How much data can flow through a network in a given amount of time, often characterized by the question "How big is the pipe?"

Node. A computer connected to a network.

Operating system. The program that manages the computer's resources for the use of other programs, launches other computer programs, and usually manages the interface with the user.

Peripheral device. A piece of equipment that enables users to send or receive data to and from a computer, such as a floppy disk, printer, modem, mouse, keyboard, or monitor.

Protocol. A set of rules defining how to accomplish a task, for instance, a networking protocol defines how two nodes are supposed to communicate.

Random access memory (RAM). The main working memory of a computer, with individual memory locations that can be directly accessed at any time, unlike a sequential storage device such as a magnetic tape.

Read-only memory (ROM). A memory chip that can only be read from, not written to.

Real-time system. A system that must respond to events in the physical world within a given amount of time. For example, a heart monitor must alert hospital personnel within fractions of a second, not minutes.

Relay. An electromagnetic switch.

Software. Something that you cannot touch, as opposed to hardware, which you can touch. Software is executable programs, not data.

Vacuum tube. A glass tube holding a vacuum and containing an electronic switch or amplifier.

Wide area network (WAN). Network allowing computers to communicate across the street or around the world. *Also see* the definition of Local area network (LAN).

Bibliography

Abbate, Janet. *Inventing the Internet.* Cambridge, MA: MIT Press, 2000.

Atanasoff, John V. "Advent of Electronic Digital Computing." *Annals of the History of Computing* 6, no. 3 (July 1984): 229–282.

Backus, John. "The History of Fortran I, II, and III." *IEEE Annals of the History of Computing* 20, no. 4 (1998): 68–78.

Bashe, Charles J., Lyle R. Johnson, John H. Palmer, and Emerson W. Pugh. *IBM's Early Computers.* Cambridge, MA: MIT Press, 1986.

Berners-Lee, Tim. *Weaving the Web: The Original Design and Ultimate Destiny of the World Wide Web by Its Inventor.* San Francisco: HarperSanFrancisco, 1999.

Burks, Alice R., and Arthur W. Burks. *The First Electronic Computer: The Atanasoff Story.* Ann Arbor: University of Michigan Press, 1988.

Campbell-Kelly, Martin. *From Airline Reservation to Sonic the Hedgehog: A History of the Software Industry.* Cambridge, MA: MIT Press, 2003.

Campbell-Kelly, Martin, and William Aspray. *Computer: A History of the Information Machine.* New York: Basic Books, 1996.

Ceruzzi, Paul E. *A History of Modern Computing.* Cambridge, MA: MIT Press, 1998.

Chandler, Alfred D., Jr. *Inventing the Electronic Century: The Epic Story of the Consumer Electronics and Computer Industries.* New York: Free Press, 2001.

Crevier, Daniel. *AI: The Tumultuous History of the Search for Artificial Intelligence.* New York: Basic Books, 1993.

Cringely, Robert X. *Accidental Empires: How the Boys of Silicon Valley Make Their Millions, Battle Foreign Competition, and Still Can't Get a Date.* Reading, MA: Addison Wesley, 1992.

Flowers, Thomas H. "The Design of the Colossus." *Annals of the History of Computing* 5, no. 3 (July 1983): 239–252.

Free as in Freedom: Richard Stallman's Crusade for Free Software. http://www.faifzilla.org/. Accessed August 25, 2004.

Freiberger, Paul, and Michael Swaine. *Fire in the Valley: The Making of the Personal Computer.* 2nd ed. New York: McGraw-Hill, 2000.

Gillies, James, and Robert Cailliau. *How the Web Was Born: The Story of the World Wide Web.* Oxford: Oxford University Press, 2000.

Hafner, Katie, and Matthew Lyon. *Where Wizards Stay Up Late: The Origins of the Internet.* New York: Simon & Schuster, 1996.

Hanson, Dirk. *The New Alchemists.* Boston: Little, Brown and Company, 1982.

Hiltzik, Michael A. *Dealers of Lightning: Xerox PARC and the Dawn of the Computer Age.* New York: HarperBusiness, 1999.

Hinsley, Francis H., and Alan Stripp, eds. *Codebreakers: The Inside Story of Bletchley Park.* Oxford: Oxford University Press, 1993.

Hodges, Andrew. *Alan Turing: The Enigma.* New York: Simon & Schuster, 1983.

Hogan, James P. *Mind Matters: Exploring the World of Artificial Intelligence.* New York: Ballantine, 1997.

Hsu, Feng-hsiung. *Behind Deep Blue: Building the Computer That Defeated the World Chess Champion.* Princeton, NJ: Princeton University Press, 2002.

Ifrah, Georges. *The Universal History of Computing: From the Abacus to the Quantum Computer.* New York: John Wiley & Sons, 2001.

Internet Society (ISOC): Histories of the Internet. http://www.isoc.org/internet/history/. Accessed July 6, 2004.

Johnson, Luanne (James). "A View from the 1960s: How the Software Industry Began." *IEEE Annals of the History of Computing* 20, no. 1 (1998): 36–42.

Kent, Steven L. *The Ultimate History of Video Games.* Prima Publishing, 2001.

Kubie, Elmer C. "Recollections of the First Software Company." *IEEE Annals of the History of Computing* 16, no. 2 (1994): 65–71.

Kurtz, Thomas E. "BASIC Session." In *History of Programming Languages*, edited by Richard L. Wexelblat, 515–549. New York: Academic Press, 1981.

Lohr, Steve. *Go To: The Story of the Math Majors, Bridge Players, Engineers, Chess Wizards, Maverick Scientists and Iconoclasts—The Programmers Who Created the Software Revolution.* New York: Basic Books, 2001.

Macrae, Norman. *John von Neumann.* New York: Pantheon Books, 1992.

McCartney, Scott. *ENIAC: The Triumphs and Tragedies of the World's First Computer.* New York: Walker and Company, 1999.

Microsoft Museum Home Page. http://www.microsoft.com/museum/default.mspx. Accessed August 25, 2004.

Mollenhoff, Clark R. *Atanasoff: Forgotten Father of the Computer.* Ames: Iowa State University Press, 1988.

Moore, Gordon E. "Intel: Memories and the Microprocessor." *Daedalus* 125, no. 2 (1996): 55–80.

Moschovitis, Christos J.P., Hilary Poole, Tami Schuyler, and Theresa M. Senft. *History of the Internet: A Chronology, 1943 to the Present*. Santa Barbara, CA: ABC-CLIO, 1999.

Narins, Brigham, ed. *World of Computer Science*. 2 vols. Detroit, MI: Gale Group/ Thomson Learning, 2002.

Paul Baran Oral History. http://www.ieee.org/organizations/history_center/ oral_histories/transcripts/baran.html. Accessed March 14, 2004.

Queisser, Hans. *The Conquest of the Microchip: Science and Business in the Silicon Age*. Cambridge, MA: Harvard University Press, 1988.

Raymond, Eric S. *The Cathedral and the Bazaar: Musings on Linux and Open Source by an Accidental Revolutionary*. Rev. ed. Cambridge, MA: O'Reilly & Associates, 2001.

Redmond, Kent C., and Thomas M. Smith. *From Whirlwind to MITRE: The R&D Story of the SAGE Air Defense Computer*. Cambridge, MA: MIT Press, 2000.

Reid, T. R. *The Chip: How Two Americans Invented the Microchip and Launched a Revolution*. New York: Random House, 2001.

Riordan, Michael, and Lillian Hoddeson. *Crystal Fire: The Birth of the Information Age*. New York: W. W. Norton, 1997.

Rojas, Raúl, and Ulf Hashagen. *The First Computers: History and Architectures*. Cambridge, MA: MIT Press, 2000.

Segaller, Stephen. *Nerds 2.0.1: A Brief History of the Internet*. New York: TV Books, 1988.

Siegfried, Tom. *The Bit and the Pendulum*. New York: John Wiley & Sons, 2000.

Slater, Robert. *Portraits in Silicon*. Cambridge, MA: MIT Press, 1987.

Smith, Thomas M. "Project Whirlwind: An Unorthodox Development Project." *Technology and Culture* 17, no. 3: 447–464.

Stern, Nancy. *From ENIAC to UNIVAC: An Appraisal of the Eckert-Mauchly Computers*. Bedford, MA: Digital Press, 1981.

Wilkes, Maurice V. *Memoirs of a Computer Pioneer*. Cambridge, MA: MIT Press, 1985.

Williams, Michael R. *A History of Computing Technology*. 2nd ed. Los Alamitos, CA: IEEE Computer Society Press, 1997.

Yates, JoAnne. "Application Software for Insurance in the 1960s and Early 1970s." *Business and Economic History* 24, no. 1 (fall 1995): 123–134.

Index

About the Authors

ERIC G. SWEDIN is an assistant professor in Information Systems and Technologies at Weber State University. He is also a historian and published novelist.

DAVID L. FERRO is an assistant professor in Computer Science at Weber State University. He specializes in internet programming, human-computer usability, and computing culture and history.